Student's Solutions Manual

Barbara A. Brown
Anoka-Ramsey Community College

Essentials of Geometry for College Students

Second Edition

Margaret L. Lial
American River College

Barbara A. Brown
Anoka-Ramsey Community College

Arnold R. Steffensen
Northern Arizona University

L. Murphy Johnson
Northern Arizona University

Boston   San Francisco   New York
London   Toronto   Sydney   Tokyo   Singapore   Madrid
Mexico City   Munich   Paris   Cape Town   Hong Kong   Montreal

Reproduced by Pearson Addison-Wesley from electronic files supplied by the author.

Copyright © 2004 Pearson Education, Inc.
Publishing as Pearson Addison-Wesley, 75 Arlington Street, Boston, MA 02116

ISBN   0-321-17353-8

1 2 3 4 5 6 QEP 06 05 04 03

PEARSON
Addison
Wesley

CONTENTS

TO THE STUDENT

This Student's Solutions Manual is written to accompany *Essentials of Geometry for College Students,* Second Edition by Margaret L. Lial, Barbara A. Brown, Arnold R. Steffensen, and L. Murphy Johnson. It contains complete worked-out solutions to the odd-numbered exercises and to all chapter review exercises, practice tests, and cumulative review exercises. Worked-out solutions to the practice exercises in each section are also provided.

To make the best use of this guide as an aid to learning the material, consider the following procedure.

1. As you study each section in the text, read through every example and make sure you understand each step. This may sometimes require more than just one reading. Note any questions you have in the margin of the text, and ask them during class discussions.

2. As you read through the examples, work any practice exercises included in the section discussion. If you have trouble or want to check your work on these exercises, refer to the solutions given in this manual.

3. After you have completed the material in the body of each section, you must apply what you have learned by working the exercises. You may check your results with the step-by-step solutions given here.

4. After you have completed each chapter in the text, read and study the chapter review material which includes key terms and symbols and proof techniques. Then work the chapter review exercises. Solutions to all review problems are included in this manual if you have difficulty. Finally, you may test your understanding of the material in the chapter by taking the practice test found in the text: again, complete solutions are given in this manual.

5. To help you study for a midterm or final exam, use the cumulative reviews found at the end of chapters 3, 7, and 10. Complete solutions are given in this manual.

By following these steps, you will get the most out of this guide and will also improve your chance of success in the course.

The following people have made valuable contributions to the production of this *Student's Solutions Manual*: Judy Martinez, typist; Sheri Minkner, artist and typist; Sarah Kueffer and Bettie Truitt, accuracy checkers.

CHAPTER 1 FOUNDATIONS OF GEOMETRY

Section 1.1 Inductive and Deductive Reasoning

1.1 PRACTICE EXERCISES

1. In the list of numbers

$$1, 4, 9, 16, 25$$

we might observe that each is a perfect square. That is, $1 = 1^2, 4 = 2^2, 9 = 3^2, 16 = 4^2$, and $25 = 5^2$. As a result, we might conclude that the next text term is $36 = 6^2$.

(b) If the number 1, 2, 3, 4, and 5 are each placed next to it's mirror image, the elements in the list would result. Thus, if we write 6 next to its mirror image, 36, this is the next element in the list.

2. (a) The conclusion follows using deductive reasoning with the given premise "This year is leap year" and the accepted but unstated well-known premise "Leap year occurs every four years."

(b) Inductive reasoning is used and does not result in a correct conclusion. Just because the Celtics won the championship in all years except 1967 does not mean they won it in 1967 too. In fact, in 1967, the Philadelphia 76ers won the title.

1.1 SECTION EXERCISES

1. Since each number in the list (after the first one) can be found by adding 5 to the preceding one, the next number is $23 + 5 = 28$

3. Since each number in the list (after the first one) can be found by multiplying the preceding one by 3, the next number is $(3)(81) = 243$

5. Since each number in the list (after the first one) can be found by multiplying the preceding one by -2, the next number is $(-2)(16) = -32$

7. Each number in the list (after the first two) can be obtained by adding the two preceding numbers. Thus, the next number is $13 + 21 = 34$

9. We might note that these are the first letters in the seven days of the week. The next letter would be S corresponding to Saturday.

11. We might note that these are the last five letters of the alphabet. Thus, the next letter would be U.

13. From the examples, we can reason the sum of two odd numbers is an even number.

15. A postulate is a statement that is assumed true without proof. A definition is a statement that gives us a new term to be used in the system.

17. A postulate is a statement that is assumed true, and a theorem is a statement that is proved.

19. The hypothesis follows "if"; the conclusion follows "then."

21. The statement is a postulate (it is assumed true.)

23. Since the countries have different postulates, it is unlikely that they will have the same theorems. That is, there will probably be disagreement in their negotiations.

25. The conclusion does not follow; this is a fallacy in reasoning.

27. The conclusion follows logically; deductive reasoning.

29. The conclusion does not follow; inductive reasoning.

31. The conclusion follows logically; deductive reasoning.

33. The conclusion follows logically; deductive reasoning.

35. We cannot conclude it: is necessarily green, only that *if* it hops or is a frog *then* it is green.

37. The older dog is the younger dog's father.

39. 12 (Since *all* but 12 died, 12 were left alive.)

41. It would have been impossible in 300 B.C. to know that that was the date as we now know it.

43. 9 minutes; it then doubles the next minute to be full at 10 minutes.

45. Once; the next time you are subtracting 5 from 20 *not* 25.

47. Identify the sacks as sack 1, sack 2, and sack 3. Take one coin from sack 1, two coins from sack 2 and three coins from sack 3. Place the coins on the scale. The reading will be 6 lb 1 oz, 6 lb 2 oz, or 6 lb 3 oz. The number of ounces identifies the sack with the counterfeit coins.

49. The shortest route for the bug to take is shown in the figure below. The fact that it is the shortest is easy to see if we "unfold" the top face of the cube and draw a straight line between points *A* and *B*. (The shortest distance between two points is the distance along the straight line joining them.)

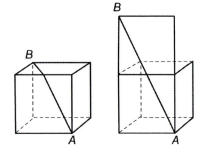

51. The four pieces could be arranged as shown below. The fallacy here is that the pieces do not actually fit to form the diagonal line as shown. Some parts overlap covering part of the area and giving rise to the "missing" little square.

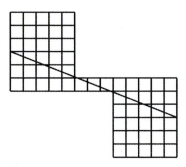

53. Take away the seven matches forming the F and the E which leaves IV, the Roman numeral for four.

55. To keep track of the information given, we can use two grids and agree to fill the positions in each grid with an "x" or an "o" to indicate that the statement is true or false, respectively.

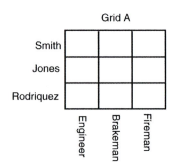

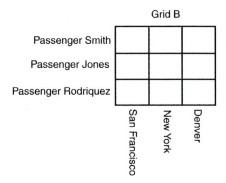

Step 1 Postulate (d) allows us to fill Grid B as follows.

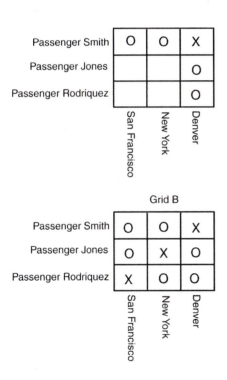

	San Francisco	New York	Denver
Passenger Smith	O		
Passenger Jones	O		
Passenger Rodriquez	X	O	O

Step 2 Postulates (c) and (g) give that the physicist lives in Denver, so he is not Passenger Rodriquez who lives in San Francisco and not Passenger Jones, who by Postulate (e) forgot his algebra; so he must be Passenger Smith. Thus, GRID B can be filled in making it complete.

	San Francisco	New York	Denver
Passenger Smith	O	O	X
Passenger Jones			O
Passenger Rodriquez			O

Grid B

	San Francisco	New York	Denver
Passenger Smith	O	O	X
Passenger Jones	O	X	O
Passenger Rodriquez	X	O	O

Step 3 Postulate (f) with GRID B gives the brakeman's name as Jones. Thus, GRID A becomes:

	Engineer	Brakeman	Fireman
Smith		O	
Jones	O	X	O
Rodriquez		O	

Step 4 Finally, Postulate (h) says Smith is not the fireman making GRID A look like:

	Engineer	Brakeman	Fireman
Smith		O	O
Jones	O	X	O
Rodriquez		O	

This allows us to complete Grid A as:

Grid A

	Engineer	Brakeman	Fireman
Smith	X	O	O
Jones	O	X	O
Rodriquez	O	O	X

Thus, Smith is the engineer, Jones is the Brakeman, and Rodriquez is the fireman. Passenger Smith lives in Denver,

Passenger Jones lives in New York, and

Passenger Rodriquez lives in San Francisco.

Section 1.2 Points, Lines, and Planes

1.2 PRACTICE EXERCISES

1. Since A is in the position of 3 on the number line, the coordinate of A is 3. Since B is in the position of -2 on the number line, the coordinate of B is -2. Since C is half way between -1 and 0, the coordinate of C is $-\frac{1}{2}$. Since D is about $\frac{3}{4}$ of the way from 1 to 2, the coordinate of D is $1 + \frac{3}{4} = \frac{7}{4}$.

1.2 SECTION EXERCISES

1. Exactly one

3. C is in plane P

5. Postulate 1.1

7. Postulate 1.4

9.

11. A: 2; B: -3; C: 3.5; D: $-\frac{1}{2}$

13. 5   15. 7   17. a

19. False; The statement is true by the transitive law.

21. False; It's true by the multiplication-division postulate

23. 2. Add.-Sub Post.   4. Add-Subt Post
    6. Mult.-Div. Post.   8. Symmetric Law
    9. Substitution Law

25. Yes, if the given point is a common point where two lines meet. If five lines meet at one point, then the point exists on all five lines.

27. Yes, the line is where the two planes meet. If all five planes meet in one common place, this will be the described line.

29. By Postulate 1.2, three distinct points (the ends of the legs) all lie in the same plane (the floor on which the stool stands). However, four points (the ends of a 4-legged stool) need not all lie in the same plane (the floor on which that stool stands).

Section 1.3 Segments, Rays, and Planes

1.3 PRACTICE EXERCISES

1. By the extension of the segment addition postulate, the three segments of length $x + 2$, $2x + 1$, and x must add to equal 23. Thus, we must solve:

$$(x+2)+(2x+1)+x = 23$$
$$x+2+2x+1+x = 23 \quad \text{Clear parentheses}$$
$$4x+3 = 23 \quad \text{Combine terms}$$
$$4x = 20 \quad \text{Subtract 3 from both sides}$$
$$x = \frac{20}{4} \quad \text{Divide both sides by 4}$$
$$x = 5 \quad \text{Simplify}$$

2. First sketch a figure showing Z on $\overleftrightarrow{XY}$ but not on $\overrightarrow{XY}$

(a) Then Z is on $\overrightarrow{YX}$ but not on $\overrightarrow{XY}$

(b) Z is not on the segment $\overline{YX}$ which includes only points between X and Y along with X and Y.

(c) Then Y is on $\overrightarrow{ZX}$ but not on $\overrightarrow{XZ}$.

(d) $\overrightarrow{XY}$ only has one endpoint (like every ray) which is X.

(e) The endpoints of the segment $\overline{XY}$ are X and Y (which is true for any segment, a segment has two endpoints).

(f) $\overrightarrow{XY}$ and $\overrightarrow{YX}$ are different rays. Notice that Z is on $\overrightarrow{YX}$ but not on $\overrightarrow{XY}$.

(g) Since $ZY = ZX + XY$ by the segment addition postulate, substituting we have:

$$ZY = 3 + 5 = 8 \text{ cm}$$

3. (a) If $m\angle A = 27°$ and $\angle A$ and $\angle B$ are complementary, then

$$m\angle A + m\angle B = 90°.$$

Substitute $27°$ for $m\angle A$ and solve for $m\angle B$.

$$27° + m\angle B = 90°$$
$$m\angle B = 90° - 27°$$
$$m\angle B = 63°$$

(b) If $m\angle P = 74°26'52''$ and $\angle P$ and $\angle Q$ are supplementary, then

$$m\angle P + m\angle Q = 180°.$$

Substitute $74°26'52''$ for $m\angle P$ and solve for $\angle Q$.

$$74°26'52'' + m\angle Q = 180°$$
$$m\angle Q = 180° - 74°26'52''$$

To find $m\angle Q$ we must "borrow" minutes and seconds from degrees as follows.

$$180° = 179°60' = 179°59'60''$$

Then

$$180° - 74°26'52'' = 179°59'60'' - 74°26'52''$$
$$= (179° - 74°) + (59' - 26') + (60'' - 52'')$$
$$= 105° + 33' + 8''$$
$$= 105°33'8''$$

4. If $\angle P$ and $\angle Q$ are supplementary then

$$m\angle P + m\angle Q = 180°$$

Substitute $(2y - 9)°$ for $m\angle P$ and $(7y)°$ for $m\angle Q$ and solve for y.

$$(2y - 9)° + (7y)° = 180°$$
$$(2y - 9) + (7y) = 180$$
$$2y - 9 + 7y = 180$$
$$9y - 9 = 180$$
$$9y = 189$$
$$y = 21$$

1.3 SECTION EXERCISES

1. True

3. True

5. True

7. False (the rays are opposite in direction)

9. True

11. False (a line has no end points)

13. True

15. False ($\angle ACB$ is a straight angle)

17. True

19. True

21. $65°$

23. $90°$

25. $115°$

27. $72°$ (subtract $18°$ from $90°$)

29. $53°20'$ (subtract $36°40'$ from $90°$)

31. $106°$ (subtract $74°$ from $180°$)

33. $122°25'$ (subtract $57°35'$ from $180°$)

35. $45°$

37. $40°$

39. $25°$

41. $60°$ (solve $180 - x = 4(90 - x)$)

43. Acute

45. Straight

47. Acute

49. Acute

51. Right

53. 107°10′ (add the angles)

55. 139°20′16″ (add the angles)

57. Two

59. Two

61. It depends on the length of $\overline{AB}$. If $AB < 5$ there is only one. If $AB \geq 5$ there are two.

63. 45°

65. $x = 9$ (solve $(25 - x) + x + 3x = 52$)

67. $y = 20$ (solve $(30 - y) + (9y - 10) = 180$)

Section 1.4 Introduction to Deductive Proofs

1. *Given:* The President gets his budget passed.

Prove: He will be voted out of office.

1. Given 2. Taxes will rise. 3. Premise 1 4. They will go to the polls. 5. He will be voted out of office.   ∴ If the President gets his budget passed, then he will be voted out of office.

3. *Given:* I watch TV.

Prove: My parents will be upset.

Proof: STATEMENTS   REASONS

　　　1. I watch TV.   1. Given

　　　2. I will not do my homework.   2. Premise 1

　　　3. I will fail geology.   3. Premise 3

　　　4. I will lose my scholarship   4. Premise 2

　　　5. My parents will be upset.   5. Premise 4
　　　∴ If I watch TV, then my parents will be upset.

5. No. What we proved is *If, I watch TV, then* we know that my parents will be upset.

7. No, the theorem could not be proved.

9. If I am watching football, then it is Sunday.

11. If you go to the airport, then you will fly in an airplane.

13. The city is not large.

15. Joe ran in the race.

17. If this animal is not a bird, then it does not have two legs.

19. If it isn't gold, then it doesn't glitter.

21. If a figure is not a parallelogram, then it is not a rectangle.

23. If we are not the conference champions, then we lost to Central State (excluding a tie).

25. hypothesis: two lines have slopes m_1 and m_2 where $m_1 = m_2$

conclusion: the lines are parallel.

27. hypothesis: two angles are vertical angles

conclusion: the angles are congruent

Section 1.5 Formalizing Geometric Proofs

1. 1. Given 2. Given 3. Given 4. Given

5. Addition-Subtraction Post. (Post. 1.9)

6. Segment Addition Postulate (Post. 1.13)

7. Substitution Law (Post. 1.11)

3. 1. $\angle ABC$ is a right angle 2. Def. of right angle

3. Angle Addition Postulate (Post. 1.14)

4. Transitive Law (Post. 1.8) 5. Given

6. Def. of complementary angles

7. Symmetric and Transitive Laws using statements 4 and 6

8. $m\angle ABD = m\angle 1$

5. $\angle 1$ and $\angle 2$; $\angle 2$ and $\angle 3$; $\angle 3$ and $\angle 4$; $\angle 4$ and $\angle 1$

7. 150°(the supplement of 30°)

9. 150°(the supplement of 30°)

11. 15° $m\angle 3 = 135°$ and $m\angle 5 = 30°$; $m\angle 6 = 180° - 165° = 15°$)

13. Yes, by Theorem 1.11

15. Yes, if they are the same measure; which must be 90°.

17. No, obtuse means greater than 90°. The sum of two angles greater than 90 $\neq$ 180.

19. Yes, obtuse means greater than 90°. An example is 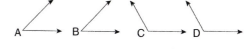 $m\angle 1 \cong m\angle 2$ and they are obtuse.

21. Yes, the lines must be perpendicular.

23. No, vertical angles must be congruent.

25. Yes, adjacent means they have a common vertex and a common side. These angles can be supplementary if their sum is 180°.

27. *Given*: $\angle A$ and $\angle C$ are supplementary
$\angle B$ and $\angle D$ are supplementary
$m\angle C = m\angle D$

Prove: $m\angle A = m\angle B$

Proof:

	STATEMENTS		REASONS
1.	$\angle A$ and $\angle C$ are supp.	1.	Given
2.	$m\angle A + m\angle C = 180°$	2.	Def. of supp. $\angle$
3.	$\angle B$ and $\angle D$ are supp.	3.	Given
4.	$m\angle B + m\angle D = 180°$	4.	Def. of supp. $\angle$
5.	$m\angle A + m\angle C = m\angle B + m\angle D$	5.	Sym. and Trans. Laws
6.	$m\angle C = m\angle D$	6.	Given
7.	$m\angle A = m\angle B$	7.	Add.-Subt. Post.

29. *Given: PR = QS*

Prove: PQ = RS

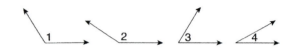

STATEMENTS	REASONS
1. $PR = QS$	1. Given
2. $QR = QR$	2. Ref. Law
3. $PR - QR = QS - QR$	3. Add.-Subt. Law
4. $PQ = PR - QR$	4. Seg.-Add. Post
5. $RS = QS - QR$	5. Seg.-Add. Post.
6. $PQ = RS$	6. Substitution Law

31. *Given*: $\angle 3$ and $\angle 4$ are complementary

$\angle 1$ and $\angle 3$ are supplementary

$\angle 2$ and $\angle 4$ are supplementary

Prove: $m\angle 1 + m\angle 2 = 270°$

Proof:

STATEMENTS	REASON
1. $\angle 3$ and $\angle 4$ are complementary	1. Given
2. $m\angle 3 + m\angle 4 = 90°$	2. Def. of comp. $\angle$
3. $\angle 1$ and $\angle 3$ are supplementary	3. Given
4. $m\angle 1 + m\angle 3 = 180°$	4. Def. of supp. $\angle$
5. $\angle 2$ and $\angle 4$ are supplementary	5. Given
6. $m\angle 2 + m\angle 4 = 180°$	6. Def. of supp. $\angle$.
7. $m\angle 1 + m\angle 2 + m\angle 3 + m\angle 4 = 360°$	7. Add.-Subt. Post. using statements 4 and 6
8. $m\angle 1 + m\angle 2 + 90° = 360°$	8. Substitution Law using statements 2
9. $m\angle 1 + m\angle 2 = 270°$	9. Add.-Subt. Post

Note to students about proofs: Proofs are not unique. Your proof may differ slightly from the solutions manual. This does not mean your proof is necessarily incorrect. Consult with your instructor.

Section 1.6 Constructions Involving Lines and Angles

1. Use Construction 1.1.

3. Use Construction 1.2.

5. Use Construction 1.2 twice.

7. Use Construction 1.3.

9. Use Construction 1.3.

11. Use Construction 1.4.

13. Use Construction 1.6.

15. Use Construction 1.6 three times.

17. A ruler is used to measure lengths, but a straight edge is used only to draw a straight line between two points.

19. Exactly one

21. Use Construction 1.5.

23. Construct the bisector of the angle with vertex at the bridge and sides containing the ranger station and the cabin. Locate the point of intersection of this bisector (ray) and the edge of the forest.

25. 1. B is the midpoint of $\overline{AC}$ 2. Def. of midpoint 3. Segment Addition Postulate (Post. 1.13) 4. Substitution Law (Post. 1.11) 5. $2AB = AC$ 6. Multiplication-Division Law (Post. 1.10)

27. 1. B is the midpoint of $\overline{AC}$ 3. Given 4. $PQ = \dfrac{PR}{2}$ 5. $AC = PR$ 6. Multiplication-Division Law (Post. 1.11)

29. The angles are complementary. Adjacent angles have a common vertex and a common side. If the non-shared sides are perpendicular this means the sum of the adjacent angles is $90°$, thus the angles are complementary.

Chapter 1 Review Exercises

1. The four parts to any axiomatic system are undefined terms, definitions, axioms or postulates, and theorems.

2. They serve as a starting point when building the system.

3. Since each number in the list (after the first one) can be obtained by adding 5 to the preceding one, the next number is $12 + 5 = 17$.

4. These letters form every other letter in the alphabet. Thus, the next letter would be two letters after I, which is K.

5. Since each number in the list (after the first one) can be obtained by multiplying the preceding one by $\frac{1}{2}$, the next number is $(\frac{1}{2})(\frac{1}{16}) = \frac{1}{32}$.

6. The conclusion follows logically; deductive reasoning.

7. The conclusion does not follow; this is a fallacy of reasoning

8. The conclusion does not follow; inductive reasoning.

9. The conclusion follows logically; deductive reasoning

10. We cannot conclude that tomorrow is a holiday, only that *if* it is Sunday, *then* tomorrow is a holiday.

11. 3 minutes   12. 3

13. True

14. False (only if they are not on the same line)

15. True   16. True

17.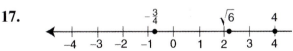

18. Reflexive Law

19. Symmetric Law

20. Transitive Law

21. Multiplication-Division Law

22. Substitution Law

23. False ($\overrightarrow{BA}$ starts at point B)

24. True

25. True

26. True

27. False ($\angle EBD$ is a straight angle)

28. True

29. False ($\angle 1$ and $\angle 2$ are vertical angles)

30. False ($\overleftrightarrow{ED}$ has no endpoints)

31. True

32. False (they are in opposite directions)

33. True

34. False ($\angle 2$ is acute)

35. $65°29'15''$ (subtract from $90°$)

36. $43°17'9''$ (subtract from $180°$)

37. $20°$

38. No (it is obtuse)

39. Exactly one

40. Two

41. *Given:* It is Gleep.

    Prove: It is Grob.

    Proof:

STATEMENTS	REASONS
1. It is Gleep.	1. Given
2. It is Glop.	2. Premise 1
3. It is Gunk.	3. Premise 3
4. It is Grob.	4. Premise 2
∴ If it is Gleep, then it is Grob.	

42. *Given:* It is a tree.

    Prove: It can be destroyed by fire.

    Proof:

STATEMENTS	REASONS
1. It is a tree.	1. Given
2. It is made of wood.	2. Premise 2
3. It will burn.	3. Premise 3
4. It can be destroyed by fire.	4. Premise 1
∴ If it is a tree, it can be destroyed by fire.	

43. If it is cold, then it is ice.

44. If you want the best for someone, then you love that person.

45. My car is not red.

46. The moon is made of green cheese.

47. If the runner does not win the race, then she is not in excellent condition.

48. If a painting is not a Picasso, then it is not valuable.

49. If I do not climb the mountain, then the weather is not good.

50. If I drive, then I do not drink.

51. $42°$ **52.** $138°$ **53.** $17°$ **54.** $163°$

55. 1. Given 2. $m\angle 1 + m\angle 2 = 90°$
3. $\angle 1$ and $\angle 3$ are vertical angles
4. Vertical angles are equal in measure
5. $\angle 2$ and $\angle 4$ are vertical angles
6. Vertical angles are equal in measure
8. $\angle 3$ and $\angle 4$ are complementary.

56. 1. Given 2. Reflexive Law 3. Add.-Subt. Post. 4. Seg. Add Post. 5. Seg. Add Post. 6. Subst. Law

57. First bisect $\overline{PQ}$ determining midpoint T. Then use the Construction 1.1 to construct segment $\overline{CD}$ on m with length PT.

58. Use Construction 1.6.

59. Use Construction 1.3.

60. Infinitely many

61. Exactly one

62. Use Construction 1.2

Chapter 1 Practice Test

1. A postulate is a statement that is assumed true without proof; and a theorem is a statement that is proved using deductive reasoning.

2. Inductive reasoning involves reaching a general conclusion based on specific observations. Deductive reasoning involves reaching a specific conclusion based on assumed general conditions.

3. Deductive reasoning

4. No. The specific observations that lead to the conclusion may not be sufficient in number.

5. Since each number in the list (after the first one) can be obtained from the preceding one by multiplying it by $-\frac{1}{5}$, the next number is $\left(-\frac{1}{5}\right)(-1) = \frac{1}{5}$.

6. No. This inductive reasoning does not necessarily lead to the same conclusion.

7. Infinitely many

8. $y + 2$

9. True

10. True

11. False $(m\angle 2 = 180° - 65° = 115°)$

12. True

13. True

14. False
($\overrightarrow{BD}$ starts at B and does not contain A)

15. True

16. False (they are complementary)

17. $56°20'$

18.   Given: I rob a bank.

Prove: I will go to jail.

Proof: STATEMENTS   REASONS

      1.   I rob a bank.   1.   Given

      2.   I will be arrested   2.   Premise 4

      3.   I will go on trial.   3.   Premise 2

      4.   I will be convicted.   4.   Premise 1

      5.   I will go to jail.   5.   Premise 3

      ∴   If I rob a bank, then I will go to jail.

19.   If it is white, then it is milk.

20.   The road to success is not difficult.

21.   If it is not a collie, then it is not a dog.

22.   If the fruit is not ripe, then it is not picked.

23.   1. Given

    2. Adj. $\angle's$ whose noncommon sides are in line are supp.

    3. Same as 2

    4. Supp's of = $\angle's$ are = in measure (Theorem 1.8)

24.   Use Construction 1.3 to find midpoint C; then Construction 1.2 to construct the desired angle.

25.   Use Construction 1.6

26.   1. Given   2. Angle-Add. Post.   3. Angle-Add. Post.   4. Substitution   5. Reflective Law
    6. Add-Subt. Law

27.   a. midpoint   b. bisector   c. perpendicular bisector.

CHAPTER 2   TRIANGLES

Section 2.1 Classifying Triangles

2.1 PRACTICE EXERCISES

1. In $\triangle ABC$, $m\angle C = 90°$ and $AC \neq BC \neq AB \neq AC$.

   (a) Since $m\angle C = 90°$, $\angle C$ is a right angle so $\triangle ABC$ is a right triangle.

   (b) Since all three sides are of unequal length, the triangle is scalene.

   (c) Sides $\overline{AC}$ and $\overline{BC}$ include $\angle C$ since the vertex of $\angle C$, C, is on both $\overline{AC}$ and $\overline{BC}$.

   (d) $\angle B$ is opposite side $\overline{AC}$.

   (e) Side $\overline{AB}$ is included between $\angle A$ and $\angle B$.

   (f) Side $\overline{BC}$ is opposite $\angle A$.

   (g) The hypotenuse of a right triangle is the side opposite the right angle. Since $\overline{AB}$ is opposite the right angle $\angle C$, $\overline{AB}$ is the hypotenuse.

   (h) The sides of a right triangle that are not the hypotenuse are called its legs. Thus, the legs of $\triangle ABC$ are $\overline{AC}$ and $\overline{BC}$. The legs include the right angle, $\angle C$.

2. Let x be the length of each side of an equilateral triangle with perimeter 75 ft. Then
$$x + x + x = 75$$
$$3x = 75$$
$$x = 25$$

   Thus each side is 25 ft long.

2.1 SECTION EXERCISES

1. D, E, and F

3. Obtuse triangle

5. $\overline{DE}$

7. $\angle F$

9. $\overline{DE}$

11. $\overline{AB}$, $\overline{BC}$, and $\overline{AC}$

13. Equilateral triangle

15. $\overline{BC}$

17. $\angle A$

19. It has no hypotenuse since it is not a right triangle.

21. $20 + 30 + 40 = 90$; 90 cm

23. 33 in ($12.5 + 12.5 + 8 = 33$)

25. 23 ft (Solve $x + x + x = 69$.)

27. base: 15 cm, sides: 45 cm

   (Solve $x + x + \frac{1}{3}x = 105$.)

29. $\triangle AED$ and $\triangle AEB$

31. $\triangle ECD$ and $\triangle ACD$

33. True

35. False (neither side of the angle is a side of the triangle)

37. False ($\angle CED$ is not adjacent to $\angle GEF$)

39. True

41. Yes, a triangle can be scalene and obtuse. Scalene means all sides have different lengths and obtuse means the triangle has one obtuse angle.

43. Yes, a right triangle can be isosceles. A right triangle contains one right angle while an isosceles triangle has two congruent sides.

45. No, an equilateral triangle has three angles that measure 60°.

Section 2.2 Congruent Triangles

2.2 PRACTICE EXERCISES

1. We can show the desired congruence in two ways.

First, since $AC = 3$ cm and $US = 3$ cm, $\overline{AC} \cong \overline{SU}$. Also, we are given that $\overline{CB} \cong \overline{UT}$, and that $\angle C$ and $\angle U$ are right angles making $\angle C \cong \angle U$. Then $\triangle ABC = \triangle STU$ by SAS

Second, since $AB = 5$ cm and $TS = 5$ cm, $\overline{AB} \cong \overline{ST}$. Also, since $AC = 3$ cm and $US = 3$ cm, $\overline{AC} \cong \overline{SU}$, and since we are also given that $\overline{CB} \cong \overline{UT}$, $\triangle ABC \cong \triangle STU$ by SSS

2.2 SECTION EXERCISES

1. Congruent by SSS

3. Since right angles are congruent, the triangles are congruent by SAS

5. Congruent by ASA

7. 1. Given   2. $\angle 3 \cong \angle 4$
    3. $\overline{BD} \cong \overline{BD}$ 4. ASA

9. 1. Given   2. $\perp$ lines form rt. $\angle$'s
    3. $\angle ABD \cong \angle CDB$ 4. $\overline{DB} \cong \overline{BD}$
    5. Given   6. SAS

Note to students about proofs: Proofs are not unique. Your proof may differ slightly from the solutions manual. This does not mean your proof is necessarily incorrect. Consult with your instructor.

11. *Proof:*

STATEMENTS	REASONS
1. $\overline{AD}$ bisects $\overline{BE}$	1. Given
2. $\overline{BC} \cong \overline{EC}$	2. Def. of bisector
3. $\overline{BE}$ bisects $\overline{AD}$	3. Given
4. $\overline{AC} \cong \overline{DC}$	4. Def. of bisector
5. $\angle ACB \cong \angle DCE$	5. Vertical angles are $\cong$
6. $\triangle ABC \cong \triangle DEC$	6. SAS

13. *Proof:* STATEMENTS   REASONS

1. $\overline{AD} \cong \overline{BD}$ so $AD = BD$ 1. Given; Def. $\cong$ seg.
2. $\overline{AE} \cong \overline{BC}$ so $AE = BC$ 2. Given; Def. $\cong$ seg.
3. $AD - AE = BD - BC$ 3. Add.-Subt. Post.
4. $ED = AD - AE$ and $CD = BD - BC$ 4. Seg-Add. Post.
5. $ED = CD$ so $\overline{ED} \cong \overline{CD}$ 5. Substitution; Def. $\cong$ seg
6. $\angle D \cong \angle D$ 6. Reflexive Law
7. $\triangle ACD \cong \triangle BED$ 7. SAS

15. (a) Not correct because $\overline{BA} \not\cong \overline{ED}$

   (b) Correct

   (c) Not correct because $\overline{ED} \not\cong \overline{AB}$

   (d) Not correct, $\triangle BAC \cong \triangle FED$ by SAS

17. (b) Same   **(c)** Match
   (d) Match; yes   **(e)** Yes
   (f) No, no AA Theorem because the corresponding sides of two triangles may not match even if corresponding angles are congruent.

19. Since $AC = PR$, we must solve

$$x + 1 = 3x - 5.$$
$\quad 1 = 2x - 5$ Subtract x from both sides
$\quad 6 = 2x$ Add 5 to both sides
$\quad 3 = x$ Divide both sides by 2

21. Since $AC = x + 1$ and $x = 3$ (by Exercise 19), substituting 3 for x we obtain

$$AC = 3 + 1 = 4$$

23. Since $m\angle B = (100 + y)°$ and $y = 20$ (by Exercise 20), substituting 20 for y we obtain

$$m\angle B = (100 + 20)° = 120°.$$

25. $\overline{PR}$ **27.** $\overline{QR}$ **29.** $\angle P$

Section 2.3 Proofs Involving Congruence

1. 9 cm   **3.** 46°

5. 1. Given   2. $\angle 1 \cong \angle 2$ 3. Adj. $\angle$'s whose noncommon sides are in a line are supp. $\angle$'s
    4. $\angle 2$ and $\angle ECB$ are supplementary
    5. Supp. of $\cong$ $\angle$'s are $\cong$ 6. Reflexive Law
    7. SAS   8. $\angle G \cong \angle E$

7. 1. Given   2. Given   3. $\overline{AB} \cong \overline{CD}$; $AB = CD$
    4. Seg.-Add. Post.   5. $BD = CD + BC$
    7. Substitution law; Def. $\cong$ seg.
    8. $\triangle AFC \cong DEB$ 9. CPCTC

Note to students about proofs: Proofs are not unique. Your proof may differ slightly from the solutions manual. This does not mean your proof is necessarily incorrect. Consult with your instructor.

9. *Proof:*   STATEMENTS   REASONS

1. $\overline{AC} \cong \overline{CE}$ 1. Given
2. $\overline{DC} \cong \overline{CB}$ 2. Given
3. $\angle DCE$ and $\angle BCA$ are vertical angles   3. Def. of vert. $\angle$'s
4. $\angle DCE \cong \angle BCA$ 4. Vert. $\angle$'s $\cong$
5. $\triangle DCE \cong \triangle BCA$ 5. SAS
6. $\angle A \cong \angle E$ 6. CPCTC

11. *Proof:*

STATEMENTS	REASONS
1. $\overline{BC} \cong \overline{CD}$	1. Given
2. $\angle 1 \cong \angle 2$	2. Given
3. $\overline{AC} \cong \overline{AC}$	3. Reflexive Law
4. $\triangle ABC \cong \triangle ADC$	4. SAS
5. $\overline{AB} \cong \overline{AD}$	5. CPCTC

13. *Proof:*

STATEMENTS	REASONS
1. $\angle 1 \cong \angle 2$	1. Given
2. $\angle 3 \cong \angle 4$	2. Given
3. $\overline{DB} \cong \overline{BD}$	3. Reflexive Law
4. $\triangle ABD \cong \triangle CDB$	4. ASA
5. $\angle A \cong \angle C$	5. CPCTC

15. *Proof:*

STATEMENTS	REASONS
1. $\overline{GB} \perp \overline{AF}$ and $\overline{FD} \perp \overline{GE}$	1. Given
2. $\angle 1$ and $\angle 2$ are right angles	2. $\perp$ lines form rt. $\angle$'s
3. $\angle 1 \cong \angle 2$	3. Rt. $\angle$'s are $\cong$
4. $\overline{GD} \cong \overline{FB}$ and $\overline{GB} \cong \overline{FD}$	4. Given
5. $\triangle BGF \cong \triangle DFG$	5. SAS
6. $\angle BGF \cong \angle DFG$; $m\angle BGF = m\angle DFG$	6. CPCTC; Def. $\cong \angle$'s.
7. $\angle 3 \cong \angle 4$; $m\angle 3 = m\angle 4$	7. CPCTC; Def. $\cong \angle$'s.
8. $m\angle DFG - m\angle 3 = m\angle BGF - m\angle 4$	8. Add.-Subt. Post.
9. $m\angle 5 = m\angle DFG - m\angle 3$	9. $\angle$ Add. Post.
10. $m\angle 6 = m\angle BGF - m\angle 4$	10. $\angle$ Add. Post.
11. $m\angle 5 = m\angle 6$ so $\angle 5 \cong \angle 6$	11. Substitution; Def $\cong \angle$'s.
12. $\triangle BGC \cong \triangle DFC$	12. ASA
13. $\overline{BC} \cong \overline{DC}$	13. CPCTC

17. Since $\overline{AB}$ and $\overline{DE}$ are corresponding parts of congruent triangles $\triangle ABC$ and $\triangle EDC$ (by SAS), $AB = 105$ yd (the same as DE).

19. Because the triangle is a rigid figure that cannot be distorted like a four-sided figure. This is a result of SSS since there is only one triangle possible with three given sides. The bridge cannot change shape without breaking.

21. $\overline{AD} \cong \overline{CD}$ and $\angle ADB \cong \angle CDB$ are given. $\overline{DB} \cong \overline{DB}$ by reflexive law, thus $\triangle ADB \cong \triangle CDB$ by SAS and $\overline{AB} \cong \overline{CB}$ by CPCTC.

$$8x - 3 = 6x + 1$$
$$2x = 4$$
$$x = 2$$

23. $\angle ZWY \cong \angle XYW$ and $\angle ZYW \cong \angle XWY$ are given. $\overline{WY} \cong \overline{YW}$ by reflexive law, thus $\triangle ZYW \cong \triangle XWY$ by ASA and $\overline{WZ} \cong \overline{YX}$ by CPCTC.

$$3x - 2 = 2x + 1$$
$$x = 3$$

25. $\angle BAC \cong \angle DCA$ and $\angle CAD \cong \angle ACB$ is given information, $\overline{AC} \cong \overline{CA}$ by reflexive law, thus $\triangle ABC \cong \triangle CDA$ by ASA. $\overline{AB} \cong \overline{CD}$ by CPCTC.

Section 2.4 Isosceles Triangles, Medians, Altitudes and Concurrent Lines

2.4 PRACTICE EXERCISES

1. The reason for Statement 1 is: Given

Statement 2 is: $\overline{AE} \cong \overline{DE}$

This follows since in $\triangle AED$, $\angle A \cong \angle D$ making the sides opposite $\angle A$ and $\angle D$, $\overline{AE}$ and $\overline{ED}$, congruent.

Statement 3 is: $\angle 1 \cong \angle 2$

This follows since the only other given information not already used is $\angle 1 \cong \angle 2$.

The reason for Statement 4 is: ASA

The reason for this is because $\angle 1 \cong \angle 2$, $\angle A \cong \angle D$, and the included sides $\overline{AE}$ and $\overline{DE}$ are also congruent.

The reason for Statement 5 is: CPCTC

Note that $\overline{BE}$ and $\overline{CE}$ correspond in congruent triangles $\triangle AEB$ and $\triangle DEC$.

Statement 6 is: $\triangle BCE$ is isosceles

This follows since an isosceles triangle has two congruent sides and $\overline{BE} \cong \overline{CE}$.

2. 1. $\overleftrightarrow{ED}$ because it is the perpendicular bisector of $\overline{BC}$.

2. $\overline{AD}$ because it is a median of $\triangle ABC$.

3. $\overleftrightarrow{AF}$ because it is an altitude of $\triangle ABC$.

4. $\overleftrightarrow{BG}$ because it is an angle bisector of $\angle ABC$.

2.4 SECTION EXERCISES

1. $\overline{AB} \cong \overline{CB}$

3. $\overline{WY} \cong \overline{XY}$

5. $\overline{RS} \cong \overline{TR}$

7. 1. Given   2. $\overline{AC}$ bisects $\angle BAD$

3. Def. of $\angle$ bisector   4. Reflexive law

5. $\triangle BAE \cong \triangle DAE$ 6. CPCTC

7. $\triangle BDE$ is isosceles

9. 1. Given   2. $\overline{AC} \cong \overline{AD}$ so $AC = AD$
3. Given   4. Def. of midpoint and Def. $\cong$ seg.
5. E is the midpoint of $\overline{AD}$

11. *Proof:*   STATEMENTS

   1. $\angle 1 \cong \angle 2$

   2. $\overline{BE} \cong \overline{CE}$

   3. $\angle 3 \cong \angle 4$

   4. $\angle AEB$ and $\angle DEC$ are vertical angles

   5. $\angle AEB \cong \angle DEC$

   6. $\triangle ABE \cong \triangle DCE$

   7. $\angle A \cong \angle D$

13. An altitude is a line segment from a vertex of a triangle, perpendicular to the side opposite the vertex (or an extension of that side). What distinguishes this from the other two is an altitude of a triangle may be outside the triangle and an altitude is perpendicular to a side of the triangle.
An angle bisector is a ray that separates an angle into two congruent adjacent angles. What distinguishes this from the other two is all angle bisectors of a triangle are inside the triangle and cuts the angle into two congruent angles.
A median of a triangle is a line segment joining a vertex with the midpoint of the opposite side of the triangle. What distinguishes this from the other two is all medians of a triangle are inside the triangle and the midpoint of a side of the triangle is needed.

15. True   **17.** True   **19.** False

21. The orthocenter of a triangle is the intersection of the altitudes of the triangle. From the figure we see this point is D.

6. Def. of midpoint and Def. $\cong$ seg.
7. Seg. Add. Post.   8. Substitution law
10. Mult.-Div. Post.   13. $\angle 1 \cong \angle 2$

     REASONS

1. Given

2. Sides opp $\cong$ $\angle$'s are $\cong$

3. Given

4. Def. of vert. $\angle$'s

5. Vert. $\angle$'s are $\cong$

6. ASA

7. CPCTC

23. $AF = \frac{2}{3}(AI)$
$AF = \frac{2}{3}(10.5)$
$AF = 7$
$FI = AI - AF$
$FI = 10.5 - 7$
$FI = 3.5$ inches

25. $AF = \frac{2}{3}(AF + FI)$
$3AF = 2(AF + 6.8)$
$3AF = 2AF + 13.6$
$AF = 13.6$
$AI = AF + FI$
$AI = 13.6 + 6.8$
$AI = 20.4$ m

27. By Theorem 2.12, the bisectors of the angles of a triangle meet at a point equidistant from the sides of a triangle. The incenter is the point of intersection of the angle bisectors, thus point E is the desired point.

29. No, the centroid is on a median of a triangle. The median joins a vertex with the midpoint of the opposite side. A median is always inside a triangle.

Section 2.5 Proving Right Triangles Congruent

2.5 PRACTICE EXERCISES

1. 3. Since both triangles contain right angles, they are right triangles by the definition of a right triangle.

4. Given

5. Since B is the midpoint of $\overline{AC}$, definition of a midpoint states it separates it into two congruent parts.

6. Statements 4 and 5 show two pairs of legs in the right triangles are congruent, thus the triangles are congruent by LL.

2.5 SECTION EXERCISES

1. LA

3. LA

5. $\triangle ABT$

7. Given

9. Given

11. Def. rt $\triangle$

13. Given

15. Reflexive Law

Note to students about proofs: Proofs are not unique. Your proof may differ slightly from the solutions manual. This does not mean your proof is necessarily incorrect. Consult with your instructor.

17.

STATEMENTS	REASONS
1. $\overline{XY} \cong \overline{YZ}$	1. Given
2. $\overline{WY} \cong \overline{WY}$	2. Reflexive Law
3. $\overline{WY} \perp \overline{XZ}$	3. Given
4. $\angle XYW$ and $\angle ZYW$ are rt. $\angle$'s	4. $\perp$ lines form rt $\angle$'s
5. $\triangle XYW$ and $\triangle ZYW$ are rt. $\triangle$ⓢ	5. Def. rt $\triangle$
6. $\triangle XYW \cong \triangle ZYW$	6. LL

19.

STATEMENTS	REASONS
1. $\overline{AB} \cong \overline{DE}$	1. Given
2. $\angle ACB \cong \angle ECD$	2. Vertical $\angle$'s $\cong$
3. $\overline{AB} \perp \overline{BD}$, $\overline{DE} \perp \overline{BD}$	3. Given
4. $\angle ABC$ and $\angle EDC$ are rt $\angle$'s	4. $\perp$ lines form rt $\angle$'s
5. $\triangle ABC$ and $\triangle EDC$ are rt $\triangle$ⓢ	5. Def. rt $\triangle$
6. $\triangle ABC \cong \triangle EDC$	6. LA
7. $\overline{AC} \cong \overline{EC}$	7. CPCTC

Section 2.6 Constructions Involving Triangles

2.6 PRACTICE EXERCISES

1. We use Construction 2.2. First construct an angle equal to ∠A by using Construction 1.2. Place the compass point at the vertex of this angle and mark off lengths equal to BC on each of its sides. Then draw the segment formed by these points to make the desired △DEF.

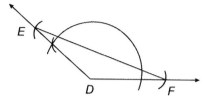

2.6 SECTION EXERCISES

1. Use Construction 2.1.

3. Use Construction 2.2.

5. Use Construction 2.3.

7. Any two obtuse angles can be tried since their noncommon sides will not intersect.

9. Use Construction 2.2.

11. The sides do not intersect to form a triangle.

13. First construct a right angle using Construction 1.4, then use Construction 2.2.

15. There are two triangles with these parts.

17. Use Construction 2.5 three times. The point of concurrency is the orthocenter.

19. Use Construction 2.6 three times. The point of concurrency is the centroid.

21. Use Construction 1.6 three times. The point of concurrency is the incenter.

23. No, the orthocenter is the point of concurrency of the altitudes of a triangle. Some altitudes in an obtuse triangle will be outside the triangle, thus, so will the orthocenter.

25. Yes, the incenter is the point of concurrency of the angle bisectors.

27. They intersect at the circumcenter.

29. Use Construction 2.4.

Chapter 2 Review Exercises

1. *A*, *B*, and *C*

2. $\overline{AB}$, $\overline{BC}$, and $\overline{CA}$

3. Scalene triangle

4. Obtuse triangle

5. $\overline{AC}$

6. ∠*B*

7. $\overline{BC}$

8. ∠*C*

9. 44 cm (44 = 12 + 16 + 16)

10. base: 13 ft, sides: 26 ft

(Solve $x + x + \frac{1}{2}x = 65$.)

11. True

12. True

13. False (∠*FEG* is not adjacent to ∠*BED*)

14. True

15. Yes

16. SSS

17. ASA

18. 1. Given   2. $\overline{AC} \cong \overline{AE}$ 3. $\angle A \cong \angle A$
    4. $\triangle ACF = \triangle AEB$

19. 1. Given   2. $\overline{AC}$ bisects $\angle BAD$
    3. $\angle BAC \cong \angle EAC$ 4. Reflexive Law
    5. $\triangle ABC = \triangle AEC$

Note to students about proofs: Proofs are not unique. Your proof may differ slightly from the solutions manual. This does not mean your proof is necessarily incorrect. Consult with your instructor.

20. *Proof:*

STATEMENTS	REASONS
1. $\overline{AD} \cong \overline{CB}$	1. Given
2. $\overline{AB} \cong \overline{CD}$	2. Given
3. $\overline{DB} \cong \overline{BD}$	3. Reflexive Law
4. $\triangle ADB \cong \triangle CBD$	4. SSS

21. *Proof:*

STATEMENTS	REASONS
1. $\overline{AC} \perp \overline{BD}$	1. Given
2. $\angle 3 \cong \angle 4$	2. $\perp$ lines form $\cong$ adj. $\angle$'s
3. $\angle 1 \cong \angle 2$	3. Given
4. $\overline{EC} \cong \overline{EC}$	4. Reflexive Law
5. $\triangle DEC \cong \triangle BEC$	5. ASA
6. $\overline{DC} \cong \overline{BC}$	6. CPCTC
7. $\overline{AC} \cong \overline{AC}$	7. Reflexive Law
8. $\triangle ABC = \triangle ADC$	8. SAS
9. $\overline{AB} \cong \overline{AD}$	9. CPCTC

22. Since $\overline{AC} \cong \overline{EF}$ by $\cong$ ⟨S⟩, we must solve

$x + 2 = 4x - 4$.
    $2 = 3x - 4$ Subtract x from both sides
    $6 = 3x$ Add 4 to both sides
    $2 = x$

23. Since $\angle E \cong \angle A$ by $\cong$ ⟨S⟩, we must solve

$y + 10 = 2y - 15$
    $10 = y - 15$ Subtract y from both sides
    $25 = y$ Add 15 to both sides

24. Since $AC = x + 2$ and $x = 2$ (from Exercise 22), substituting 2 for x we obtain

$$AC = 2 + 2 = 4.$$

25. Since $\angle A = (2y - 15)°$ and $y = 25$ (from Exercise 23), substituting 25 for y we obtain

$$\angle A = (2(25) - 15)° = (50 - 15)° = 35°.$$

26. $\overline{DF}$ corresponds to $\overline{BC}$.

27. $\angle B$ corresponds to $\angle D$.

28. 1. Given 2. $\overrightarrow{AC}$ bisects $\angle DAB$ 3. Def. of $\angle$ bisector; Def. $\cong \angle$'s   4. Given
    5. $\angle EBA \cong \angle DBE$ so $m\angle EBA = m\angle DBE$
    6. $\angle$ Add. Post.   7. Substitution law
    9. Sym. and trans. laws
    10. Mult.-Div. Post. and Def $\cong \angle$'s
    11. Reflexive law   12. ASA   13. $\overline{AC} \cong \overline{BE}$

29. *Proof:*

STATEMENTS	REASONS
1. E is the midpoint of $\overline{AC}$	1. Given
2. $\overline{AE} \cong \overline{CE}$	2. Def. of midpoint
3. E is the midpoint of $\overline{BD}$	3. Given
4. $\overline{BE} \cong \overline{DE}$	4. Def. of midpoint
5. $\angle AEB$ and $\angle CED$ are vertical angles	5. Def. of vert. $\angle$'s
6. $\angle AEB \cong \angle CED$	6. Vert. $\angle$'s are $\cong$
7. $\triangle AEB \cong \triangle CED$	7. SAS
8. $\overline{AB} \cong \overline{CD}$	8. CPCTC

30. *Proof:*

STATEMENTS	REASONS
1. $\overline{AB} \cong \overline{CD}$	1. Given
2. $\overline{BC} \cong \overline{DE}$	2. Given
3. $\angle CAE \cong \angle CEA$	3. Given
4. $\overline{AC} \cong \overline{CE}$	4. Sides opp. $\cong \angle$'s are $\cong$
5. $\triangle ABC \cong \triangle CDE$	5. SSS
6. $\angle B \cong \angle D$	6. CPCTC

31. *Proof:*

STATEMENTS	REASONS
1. $\triangle ABD$ is isosceles with base $\overline{BD}$	1. Given
2. $\overline{AB} \cong \overline{AD}$	2. Def. of isosceles $\triangle$
3. $\triangle BDE$ is isosceles with base $\overline{BD}$	3. Given
4. $\overline{BE} \cong \overline{DE}$	4. Def. of isosceles $\triangle$
5. $\overline{AE} \cong \overline{AE}$	5. Reflexive law
6. $\triangle ABE \cong \triangle ADE$	6. SSS
7. $\angle 3 \cong \angle 4$	7. CPCTC
8. $\angle 3$ and $\angle 1$ are supplementary and $\angle 4$ and $\angle 2$ are supplementary	8. Adj. $\angle$'s whose noncommon sides are in line are supp.
9. $\angle 1 \cong \angle 2$	9. Supp. of $\cong \angle$'s are $\cong$

32. *Proof:*

STATEMENTS	REASONS
1. $\overline{AB} \cong \overline{CB}$; $\overline{AD} \cong \overline{CD}$	1. Given
2. $\overline{BD} \cong \overline{BD}$	2. Reflexive law
3. $\triangle ABD \cong \triangle CBD$	3. SSS
4. $\angle ABD \cong \angle CBD$	4. CPCTC
5. $\overline{BD}$ is bisector of $\angle ABC$	5. Def. $\angle$ bisector

33. True

34. False

35. True

36. True

37. False

38. False

39.
$$CE = \frac{2}{3}(CE + EF)$$
$$17 = \frac{2}{3}(17 + EF)$$
$$3(17) = 2(17 + EF)$$
$$51 = 34 + 2(EF)$$
$$17 = 2(EF)$$
$$8.5 \text{ m} = EF$$

Since D is the midpoint of AC, $AD = CD$.
Since $AD = 12$ m, then $CD = 12$ m.
$$AC = 12 + 12 = 24 \text{ m}$$
$$AC = 24 \text{ m}$$

40. *Proof:*

STATEMENTS	REASONS
1. $\angle C \cong \angle E$	1. Given
2. $\overline{AD} \perp \overline{BC}$; $\overline{AD} \perp \overline{DE}$	2. Given
3. $\angle ABC$ and $\angle BDE$ are rt $\angle$'s	3. $\perp$ lines form rt $\angle$'s
4. $\triangle ABC$, and $\triangle BDE$ are rt $\triangle$s	4. Def. rt $\triangle$
5. B is midpoint $\overline{AD}$	5. Given
6. $\overline{AB} \cong \overline{DB}$	6. Def. of midpoint
7. $\triangle ABC \cong \triangle BDE$	7. LA

41. *Proof:*

STATEMENTS	REASONS
1. $\triangle WXY$ is isosceles triangle with $\overline{WX} \cong \overline{YX}$	1. Given
2. $\angle W \cong \angle Y$	2. Sides opp. $\cong \angle$'s are $\cong$.
3. $\overline{XZ} \cong \overline{XZ}$	3. Reflexive law
4. $\overline{WY} \perp \overline{XZ}$	4. Given
5. $\angle XZW$, $\angle XZY$ are rt $\angle$'s.	5. $\perp$ lines form rt $\angle$'s.
6. $\triangle WXZ$ and $\triangle YXZ$ are rt $\triangle$s.	6. Def. rt $\triangle$
7. $\triangle WXZ \cong \triangle YXZ$	7. LA

42. Use Construction 2.1.

43. First construct a right angle using Construction 1.4. Then use the right angle, the segment, and the acute angle with Construction 2.3.

44. Use Construction 2.6.

45. Use Construction 2.2.

46. Use Construction 2.5.

Chapter 2 Practice Test

1. Acute triangle

2. Isosceles triangle

3. $\angle A$

4. $\angle B \cong \angle C$

5. $\angle B$

6. $\overline{BC}$

7. Yes

8. Yes

9. $\angle ACD$

10. 13 cm ($AB = AC$ so $AC = 5$ cm; then $13 = 5 + 5 + 3$)

11. 33 in (all three sides measure 11 inches).

12. 28° (Since $\angle C \cong \angle D$, solve $x + 20 = 48$ giving $x = 28$.)

13. *Proof:*

	STATEMENTS	REASONS
1.	$\angle 1 \cong \angle 2$	1. Given
2.	D is the midpoint of $\overline{CE}$	2. Given
3.	$\overline{CD} \cong \overline{DE}$	3. Def. of midpoint
4.	$\overline{AC} \cong \overline{AE}$	4. Given
5.	$\angle E \cong \angle C$	5. Angles opp. $\cong$ sides are $\cong$
6.	$\triangle BCD \cong \triangle FED$	6. ASA
7.	$\overline{BD} \cong \overline{FD}$	7. CPCTC

14. *Proof:*

	STATEMENTS	REASONS
1.	$\angle 3 \cong \angle 4$	1. Given
2.	$\overline{AC} \cong \overline{AD}$	2. Sides opp. $\cong \angle$'s are $\cong$
3.	$\angle 1$ and $\angle 3$ are supplementary and $\angle 2$ and $\angle 4$ are supplementary	3. Adj. $\angle$'s whose concommon sides are in line are supp.
4.	$\angle 1 \cong \angle 2$	4. Supp. of $\cong \angle$'s are $\cong$
5.	$\overline{BC} \cong \overline{ED}$	5. Given
6.	$\triangle ABC \cong \triangle AED$	6. SAS
7.	$\angle 5 \cong \angle 6$	7. CPCTC

15. *Proof:*   STATEMENTS   REASONS

1. $\overline{AB}$ bisects $\overline{CD}$ 1. Given

2. $\overline{DE} \cong \overline{CE}$ 2. Def. bisector

3. $\angle DEB \cong \angle CEA$ 3. Vertical $\angle$'s $\cong$

4. $\angle C$ and $\angle D$ are rt. $\angle$'s   4. Given

5. $\triangle ACE$ and $\triangle BDE$ are rt $\overset{\triangle}{\text{S}}$ 5. Def. rt $\triangle$

6. $\triangle ACE \cong \triangle BDE$ 6. LA

16. Use Construction 2.2 followed by Construction 2.5 and Construction 2.6.

17. It is both an altitude and a perpendicular bisector because it is perpendicular to the base as indicated by the symbol ⌐ and it bisects the base because the marks show the segments on both sides of the bold line are congruent. It is an altitude because the segment has an endpoint at a vertex of the triangle.

18. The bold segment is a perpendicular bisector but not an altitude. It is perpendicular and bisects one side of the triangle. This segment is not an altitude because an endpoint is not at a vertex of the triangle.

19. Neither

20. (a) centroid

(b)

$$PW = \frac{2}{3}(PW + WX)$$
$$29 = \frac{2}{3}(29 + WX)$$
$$3(29) = 3 \cdot \frac{2}{3}(29 + WX)$$
$$87 = 2(29 + WX)$$
$$87 = 58 + 2(WX)$$
$$\frac{29}{2} = \frac{2(WX)}{2}$$
$$14.5 \text{ m} = WX$$

(c) $MX = 9$ m since M is the midpoint of $\overline{MN}$ (because $\overline{PX}$ is a median.)

CHAPTER 3 PARALLEL LINES AND POLYGONS

Section 3.1 Indirect Proof and the Parallel Postulate

1. 2. Assumption   3. Premise 1   4. Premise 2   5. Premise 3

3. *Given:* The weather is nice.

Prove: I am healthy.

Proof:

STATEMENTS	REASONS
1. The weather is nice.	1. Given
2. Assume I am unhealthy.	2. Assumption
3. I don't exercise.	3. Premise 2
4. I'm not playing tennis regularly.	4. Premise 1
5. The weather is bad.	5. Premise 3

But this is a contradiction of Statement 1 "the weather is nice." Thus, we must conclude that our assumption in Statement 2 was incorrect. ∴ If the weather is nice, then I am healthy.

5. Yes

7. Yes

9. No

11. No

13. One; only line ℓ

15. Use the fact that if two lines are parallel to a third line and they intersect, then there are two lines through a point that are parallel to a line not containing the point, a contradiction.

17. Yes

19. No

21. No

23. Many examples can be given.

25. The most obvious answer is the yard lines and the side lines.

27. No. If a triangle has two right angles, the uncommon sides of the angles could not intersect to form the triangle.

29. Answers will vary.

Section 3.2 Parallel Lines

3.2 PRACTICE EXERCISES

1. The reason for Statement 1 is: Given

The reason for Statement 2 is: Adj. ∠'s whose noncommon sides are in a line are supp.

Note that ∠3 and ∠2 are adjacent angles and their noncommon sides are both on line *n*.

Statement 3 is: ∠1 ≅ ∠3

Note that ∠1 and ∠3 are supplementary to the same angle, ∠2, thus are congruent.

The reason for Statement 4 is: If alt. int. ∠'s are ≅ the lines are ‖.

2. The reason for Statement 1 is: Given

The reason for Statement 2 is: If ‖ lines are cut by transv. the alt. int. are ≅ .

Note that ∠2 and ∠3 are alternate interior angles formed by transversal ℓ cutting parallel lines *m* and *n*.

The reason for Statement 3 is: Vertical ∠'s are ≅ .

Statement 4 is: ∠1 ≅ ∠2
Since ∠2 ≅ ∠3 and ∠3 ≅ ∠1, ∠2 ≅ ∠1 by the Transitive law, and so ∠1 ≅ ∠2 by the Symmetric law.

3.2 SECTION EXERCISES

1. $\angle 3$ and $\angle 5$; $\angle 4$ and $\angle 6$

3. $\angle 1$ and $\angle 5$; $\angle 2$ and $\angle 6$;
$\angle 3$ and $\angle 7$; $\angle 4$ and $\angle 8$

5. $\angle 1$, $\angle 3$, $\angle 5$, $\angle 7$

7. $\angle 4$, $\angle 6$, $\angle 8$

9. $m\angle 2 = m\angle 4 = m\angle 8 = 135°$,
$m\angle 1 = m\angle 3 = m\angle 5 = m\angle 7 = 45°$

11. Let $m\angle 1 = x$ and $m\angle 6 = 15 + 2x$
$m\angle 1 + m\angle 6 = 180°$; $x + 15 + 2x = 180$

$3x = 165$; $x = 55$

$m\angle 1 = m\angle 3 = m\angle 5 = m\angle 7 = 55°$;
$m\angle 2 = m\angle 4 = m\angle 6 = m\angle 8 = 125°$

13. $m\angle 2 = 69°$; $a \parallel c$ and $\angle 1 \cong \angle 2$ by
alt. int. $\angle$'s $\cong$

15. $m\angle 5 = 128°$; $\angle 2$ and $\angle 4$ are supplementary;
therefore $m\angle 4 = 180° - 52° = 128°$

17. $m\angle 7 = 45°$; since $m\angle 6 = 70°$ and $a \parallel b$,
$m\angle 8 = 70°$ because $\angle 6$ and $\angle 8$ are alt. int.
$\angle$'s. $m\angle 1 = 65°$ because $\angle 1$ and $\angle 3$ are
supplementary (Thm 3.10).
$m\angle 7 = 180° - m\angle 8 - m\angle 1$.
$m\angle 7 = 180° - 70° - 65° = 45°$.

19. No; Thm 3.3 says $\overleftrightarrow{BC} \parallel \overleftrightarrow{DE}$.

21. Yes; $\overleftrightarrow{BC} \nparallel \overleftrightarrow{DE}$

23. No; Thm 2.5, Transitive law and Thm 3.3
says $\overleftrightarrow{BC} \parallel \overleftrightarrow{DE}$.

25. Yes, by Thm 2.5, Transitive law, Thm 3.3

27. *Proof:*

STATEMENTS	REASONS
1. $\angle 3$ is supplementary to $\angle 4$	1. Given
2. $\angle 3$ is supplementary to $\angle 5$	2. Adj. $\angle$'s whose noncommon sides are in line are supp.
3. $\angle 4 \cong \angle 5$	3. Supp. of $\cong \angle$'s are $\cong$.
4. $m \parallel n$	4. If corr. $\angle$'s are $\cong$, then lines are $\parallel$.

29. *Proof:*

STATEMENTS	REASONS
1. $m \parallel n$	1. Given
2. $\angle ABC$ is supplementary to $\angle 5$ and $\angle BAC$ is supplementary to $\angle 1$	2. Int. $\angle$'s same side transv. are supp.
3. $\angle 5 \cong \angle 1$	3. Given
4. $\angle ABC \cong \angle BAC$	4. Supp. of $\cong \angle$'s are $\cong$
5. $\overline{BC} \cong \overline{AC}$	5. Sides opp. $\cong \angle$'s are $\cong$.
6. $\triangle ABC$ is isosceles	6. Def. of isosc. $\triangle$.

31. *Proof:*   STATEMENTS   REASONS

    1. $m \parallel n$ 1. Given

    2. $\overline{CD} \cong \overline{CE}$ 2. Given

    3. $\angle CDE \cong \angle CED$ 3. $\angle$'s opp. $\cong$ sides are $\cong$

    4. $\angle ABC \cong \angle CDE$ and   4. If lines are $\parallel$ then alt. int. $\angle$'s are $\cong$.
       $\angle BAC \cong \angle CED$

    5. $\angle ABC \cong \angle BAC$ 5. Transitive law

    6. $\overline{AC} \cong \overline{BC}$ 6. Sides opp. $\cong \angle$'s are $\cong$.

33. *Proof:*   STATEMENTS   REASONS

    1. $\overline{AB} \cong \overline{DE}$ 1. Given

    2. $\overline{AD} \cong \overline{BE}$ (Draw $\overline{AD}$ and $\overline{BE}$)   2. Given

    3. $\overline{BD} \cong \overline{BD}$ 3. Reflexive law

    4. $\triangle ABD \cong \triangle EDB$ 4. SSS

    5. $\angle ABD \cong \angle EDB$ 5. CPCTC

    6. $m \parallel n$ 6. If alt. int. $\angle$'s are $\cong$, then lines are $\parallel$.

35. *Given:* $m \parallel n$ and m and n are cut by transversal ℓ

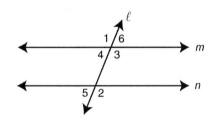

Prove: $\angle 2 \cong \angle 1$ and $\angle 5 \cong \angle 6$

Proof: STATEMENTS   REASONS

    1. $m \parallel n$ 1. Given

    2. $\angle 2 \cong \angle 3$ 2. If lines are $\parallel$, corresp. $\angle$'s $\cong$

    3. $\angle 3 \cong \angle 1$ 3. Vert. $\angle$'s $\cong$

    4. $\angle 2 \cong \angle 1$ 4. Trans. law
       Prove $\angle 5 \cong \angle 6$ in a similar manner

    5. $\angle 5 \cong \angle 4$ 5. If lines are $\parallel$, corresp. $\angle$'s $\cong$

    6. $\angle 4 \cong \angle 6$ 6. Vert. $\angle$'s $\cong$

    7. $\angle 5 \cong \angle 6$ 7. Trans. law

37. For $\overline{AB}$ to be parallel to $\overline{CD}$, $\angle BAD$ and $\angle ADC$ must be supplementary. Thus, we must solve

$$(x+50)+(x-50)=180$$
$$2x=180$$
$$x=90$$

For $\overline{AD}$ to be parallel to $\overline{BC}$, $\angle CBA$ and $\angle BAD$ must be supplementary. Thus,

$$(x+50)+y=180.$$

Since $x = 90$, we have

$$(90+50)+y=180.$$
$$140+y=180$$
$$y=40$$

39. Since $m \parallel n$, $\angle 1$ and $\angle 2$ are supplementary

$$x^2+8x=180$$
$$x^2+8x-180=0$$
$$(x-10)(x+18)=0$$
$$x-10=0 \quad x+18=0$$
$$x=10 \qquad x=-18$$

Reject $x = -18$ because it would make the measure of $\angle 1$ equal to a negative number.

Thus $x = 10$ is the solution.

41. Jose, I know since $\overline{AX} \parallel \overline{BZ}$ that $\angle XAB \cong \angle ZBC$ because they are corresponding angles. Along with the other given information, that shows $\triangle AXC \cong \triangle BZD$. By CPCTC, $\angle YCB \cong \angle ZDC$. This proves $\overline{CX} \parallel \overline{DZ}$ because using $\overline{AD}$ as the transversal if corresponding angles are congruent, lines are parallel.

Section 3.3 Polygons and Angles

3.3 PRACTICE EXERCISES

1. The sum of the measures of the angles of polygon with n sides is $S = (n-2)180$.
$$1440° = (n-2)180°$$
$$8 = n-2$$
$$10 = n$$

Thus the polygon has 10 sides.

2. The formula for the measure of each exterior angle of a regular polygon is $e = \dfrac{360°}{n}$.
$$e = \dfrac{360°}{30}$$
$$e = 12$$

Thus each exterior angle measures $12°$

3.3 SECTION EXERCISES

1. (a) triangle (b) $180°$ (c) $60°$ (d) $360°$
(e) $120°$ (f) 15 cm

3. (a) pentagon (b) $540°$ (c) $108°$ (d) $360°$
(e) $72°$ (f) 25 cm

5. (a) heptagon (b) $900°$ (c) About $128.6°$
(d) $360°$ (e) About $51.4°$ (f) 35 cm

7. (a) nonagon (b) $1260°$ (c) $140°$
(d) $360°$ (e) $40°$ (f) 45 cm

9. Solve $1620 = (n - 2)180$ for n.
$$\dfrac{1620}{180} = n-2 \quad \text{Divide both sides by 180}$$
$$9 = n-2$$
$$11 = n \qquad \text{Add 2 to both sides}$$
Thus the polygon has 11 sides.

11. Solve $2000 = (n-2)(180)$ for n.

$$\frac{2000}{180} = n - 2$$
$$11.\overline{1} = n - 2$$
$$13.\overline{1} = n$$

But n must be a whole number. Thus, there is no polygon satisfying these conditions.

13. Solve $157.5 = \dfrac{(n-2)180}{n}$ for n.

$157.5n = (n-2)180$ Multiply both sides by n

$157.5n = 180n - 360$ Distributive law

$157.5n - 180n = -360$ Subtract $180n$ from both sides

$-22.5n = -360$

$n = \dfrac{-360}{-22.5} = 16$

Thus, the polygon has 16 sides.

15. Solve $145 = \dfrac{(n-2)180}{n}$ for n.

$145n = (n-2)(180)$ Multiply both sides by n

$145n = 180n - 360$ Distributive law

$-35n = -360$ Subtract $180n$ from both sides

$n = \dfrac{-360}{-35} = 10.29$

Since n must be a whole number, there is no polygon satisfying these conditions.

17. Since the sum of the measures of the exterior angles of a triangle is $360°$, if x is the measure of the third angle, we must solve:

$$x + 200 = 360$$
$$x = 160°$$

19. Decrease

21. $60°$ (a triangle)

23. The sum of the angles of a polygon is $S = (n-2)180°$. If this sum is twice the sum of the exterior angles, $2(360°)$, we must solve:

$$(n-2)180 = 2(360)$$
$$180n - 360 = 720$$
$$180n = 1080$$
$$n = 6$$

Thus, the polygon has 6 sides.

25. The sum of the angles of a polygon is $S = (n-2)180$. The sum of the exterior angles of a polygon is $360°$. If these are equal we must solve:

$$(n-2)180 = 360$$
$$180n - 360 = 360$$
$$180n = 720$$
$$n = 4$$

Thus, the polygon has 4 sides.

27. Let x represent one of the congruent sides of the triangle. The base is represented by $x + 3$. Perimeter is the sum of the lengths of the sides.

$$x + x + x + 3 = 24$$
$$3x + 3 = 24$$
$$3x = 21$$
$$x = 7$$

The sides are 7 inches, 7 inches and 10 inches.

29. The sum of the measures of the angles of a pentagon is $(5-2)180° = 540°$

$$108 + 3(x+16) + 4(x+7) + 5x + 8 + 6(x-2) = 540$$
$$108 + 3x + 48 + 4x + 28 + 5x + 8 + 6x - 12 = 540$$
$$18x + 180 = 540$$
$$18x = 360$$
$$x = 20$$

$$m\angle A = 3(20+16) = 108°$$
$$m\angle B = 108°$$
$$m\angle C = 6(20-2) = 108°$$
$$m\angle D = 5(20) + 8 = 108°$$
$$m\angle E = 4(20+7) = 108°$$

The figure is not regular since only the angles are congruent, not the sides.

31. Consider any regular polygon with n sides and n congruent angles. By Theorem 3.17, the sum of the measures of the exterior angles of any polygon is $360°$. In problem 30, it was proven that the exterior angles of a regular polygon are congruent. Thus one exterior angle of a regular polygon with n sides and n angles measures $\frac{360°}{n}$.

33. The desired point is the point of intersection of the bisectors of $\angle DAB$ and $\angle ABC$, the incenter.

35.

Polygon	Drawing	# of sides	# of diagonals from 1 vertex
Triangle		3	0
Quadrilateral		4	1
Pentagon		5	2
Hexagon		6	3
Octagon		8	5
Decagon		10	7
Dodecagon		12	9
n-gon		n	$n-3$

Section 3.4 More Congruent Triangles

3.4 PRACTICE EXERCISE

1. 1. given   2. If lines are ∥, alt. int. $\angle$'s $\cong$ 3. $\angle ABC \cong \angle EBD$, vertical $\angle$'s $\cong$ 4. AAS

3.4 SECTION EXERCISES

1. LA   **3.** LL   **5.** HL

7. $\angle ACB \cong \angle ECD$ by vertical $\angle$'s; AAS.

9. $\overline{NL} \cong \overline{NL}$ by Reflexive law; HL

11.

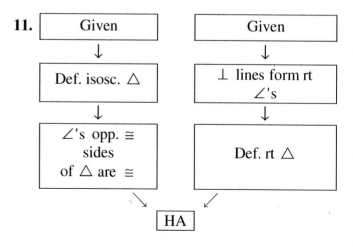

13. *Proof:*

STATEMENTS	REASONS
1. $\overline{AE} \cong \overline{BE}$; $\overline{AC} \cong \overline{BD}$ $\overline{AB} \perp \overline{CD}$	1. Given
2. $\angle AEC$ and $\angle BED$ are rt. $\angle$'s	2. $\perp$ lines form rt. $\angle$'s
3. $\triangle AEC$ and $\triangle BED$ are rt. $\triangle^{\text{(s)}}$	3. Def. rt $\triangle^{\text{(s)}}$
4. $\triangle AEC \cong \triangle BED$	4. HL

15. *Proof:*

STATEMENTS	REASONS
1. $\overline{WX} \cong \overline{ZX}$ $\angle V$ and $\angle Y$ are rt. $\angle$'s	1. Given
2. $\triangle YXZ$ and $\triangle VXW$ are rt. $\triangle^{\text{(s)}}$	2. Def. rt $\triangle$
3. $\angle YXZ \cong \angle VXW$	3. Vertical $\angle$'s $\cong$
4. $\triangle YXZ \cong \triangle VXW$	4. HA

17. *Proof:*

STATEMENTS	REASONS
1. Isos $\triangle VWX$ with base $\overline{VX}$	1. Given
2. $\overline{VW} \cong \overline{XW}$	2. Def. isos $\triangle$
3. $\overline{WY}$ is perp. bisector of $\overline{VX}$	3. Given
4. $\overline{VY} \cong \overline{XY}$	4. Def. seg. bisector
5. $\angle VYW$ and $\angle XYW$ are rt. $\angle$'s.	5. $\perp$ lines form rt. $\angle$'s
6. $\triangle VWY$ and $\triangle XWY$ are rt. $\triangle^{\text{(s)}}$	6. Def. rt $\triangle$
7. $\triangle VWY \cong \triangle XWY$	7. HL

Note: Using different statements, the $\triangle^{\text{(s)}}$ could be $\cong$ by HA or LL or LA or SAS or AAS or SSS or ASA.

19. *Proof:*

STATEMENTS	REASONS
1. $\overline{AB} \cong \overline{AD}$ $\angle ACB$ and $\angle ACD$ are rt. $\angle$'s	1. Given
2. $\triangle ACB$ and $\triangle ACD$ are rt. $\triangle^{\text{(s)}}$	2. Def. rt $\triangle$
3. $\overline{AC} \cong \overline{AC}$	3. Reflexive law
4. $\triangle ACB \cong \triangle ACD$	4. HL
5. $\angle BAC \cong \angle DAC$	5. CPCTC
6. $\overline{AC}$ bisects $\angle BAD$	6. Def. $\angle$ bisector

21.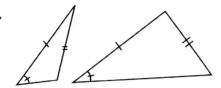

Chapter 3 Review Exercises

1. The first statement in a direct proof is P, and we form successive statements arriving at Q. The first statement in an indirect proof is $\sim Q$ and we form successive statements arriving at a contradiction (usually $\sim P$). Thus, our assumption of $\sim Q$ is wrong so we have Q, and therefore $P \to Q$.

2. *Given:* I have the money.

   Prove: I will buy my wife a present.

   Proof:

STATEMENTS	REASONS
1. I have the money	1. Given
2. Assume I don't buy my wife a present.	2. Assumption
3. My wife will be unhappy.	3. Premise 3
4. She won't wash my shirts.	4. Premise 4
5. I can't go to work.	5. Premise 1
6. I won't have any money.	6. Premise 2

   But this is a contradiction of Statement 1 "I have money." Thus, the assumption that "I don't buy my wife a present" is incorrect so we must conclude that I bought her one.
   ∴ If I have the money, then I will buy my wife a present.

3. For a given line ℓ and a point P not on ℓ, one and only one line through P is parallel to ℓ.

4. Yes, m and n are both $\perp$ to a third line, ℓ, therefore by Theorem 3.1 $m \parallel n$.

5. No, because the Parallel Postulate says there can only be one line through P parallel to ℓ.

6. Yes, since $\ell \parallel m$, alt. int. $\angle$'s $\cong$

7. No

8. Yes, $\angle 1 \cong \angle 2$ by #6 and $\angle 2 \cong \angle 3$ by vertical $\angle$'s thus $\angle 1 \cong \angle 3$ by Transitive law.

9. Yes, since $\ell \parallel m$, same side interior $\angle$'s are supp.

10. $\angle 1$, $\angle 2$, and $\angle 3$

11. 30 (Solve: $x + 20 = 3x - 40$)

12. 80 (Solve: $(y + 30) + (2y - 90) = 180$)

13. *Proof:*

STATEMENTS	REASONS
1. $\ell \parallel m$ and $\angle 1 \cong \angle 2$	1. Given
2. $\angle 3 \cong \angle 1$ and $\angle 4 \cong \angle 2$	2. Alt. int. $\angle$'s are $\cong$
3. $\angle 3 \cong \angle 4$	3. Trans. laws
4. $\overline{AB} \cong \overline{AC}$	4. Sides opp. $\cong \angle$'s are $\cong$
5. $\triangle ABC$ is isosceles	5. Def. of isos. $\triangle$

14. For ℓ and m to be parallel, interior angles on the same side of a transversal must be supplementary making

$$y + 40 = 180$$
$$y = 140,$$

and

$$(x + y) + 100 = 180$$
$$x + y = 80$$
$$x + 140 = 80$$
$$x = -60.$$

Thus, $y = 140$ and $x = -60$.

15. The sum of the angles of a hexagon is given by $S = (n-2)180°$ with $n = 6$.

$$S = (6-2)180 = 4(180) = 720°.$$

16. The measure of each angle in a regular hexagon is given by

$$a = \frac{(n-2)180°}{n}$$

when $n = 6$.

$$a = \frac{(6-2)180}{6} = \frac{720}{6} = 120°.$$

17. By Theorem 3.17, the sum of the measures of the exterior angles of a hexagon (in fact any polygon) is $360°$.

18. The measure of each exterior angle of a regular hexagon is given by

$$a = \frac{360}{n} = \frac{360}{6} = 60°.$$

19. To find the number of sides of a regular polygon if each interior angle measures $156°$ we must solve the following equation for n.

$$156 = \frac{(n-2)180}{n}$$

$156n = (n-2)180$ Multiply both sides by n

$156n = 180n - 360$ Distributive law

$-24n = -360$ Subtract $180n$ from both sides

$$n = \frac{-360}{-24} = 15$$

Thus, the polygon has 15 sides.

20. To find the number of sides of a polygon whose interior angles sum to $3600°$ we must solve the following equation for n.

$$3600 = (n-2)180$$
$$\frac{3600}{180} = n - 2 \quad \text{Divide both sides by } 180$$
$$20 = n - 2$$
$$22 = n$$

Thus the polygon has 22 sides.

21. Let x be the measure of the fourth exterior angle of a quadrilateral. The sum of the exterior angles of any quadrilateral is $360°$ so we must solve the following equation for x.

$$360 = x + 325$$
$$35 = x$$

Thus the fourth angle measures $35°$.

22. No, the interior and exterior $\angle$ at the same vertex are supp. so interior could be less than corresponding exterior.

23. No, if there were two right $\angle$'s, their sum is $180°$ but the sum of all three $\angle$'s in a $\triangle = 180°$.

24. $m\angle 1 = 60°$ since $\triangle BCD$ is rt. $\triangle$ and sum of $\angle$'s of a $\triangle = 180°$. $m\angle 2 = 120°$ since $\angle 1$ and $\angle 2$ are supplementary. $m\angle EAB = 360° - (120° + 90° + 90°) = 60°$. $\angle EAB$ is supp. $\angle 3$, therefore $m\angle 3 = 120°$.

25. In $\triangle ABC$,

$$90 + (2x+1) + 3(x-2) = 180$$
$$2x + 1 + 3x - 6 = 90$$
$$5x - 5 = 90$$
$$5x = 95$$
$$x = 19$$

Therefore

$$m\angle 1 = 2(19) + 1 = 39°$$
$$m\angle 2 = 3(19 - 2) = 51°$$

Because the measure of an exterior angle of a $\triangle$ ($\angle 3$) is equal to the sum of the nonadjacent interior angles, $m\angle 3 = 39 + 90 = 129°$.

26. $\angle B \cong \angle D$ is needed. $\overline{BC} \parallel \overline{AD}$ would make $\angle BCA \cong \angle DAC$ because they are alternate interior angles. $\overline{AC} \cong \overline{AC}$ by reflexive law. To use AAS to prove the 2 △s $\cong$, one more pair of $\cong$ $\angle$'s is needed.

27. $\overline{AB} \cong \overline{CB}$ is needed, the hypotenuses of the 2 △s. The legs $\overline{AD}$ and $\overline{DC}$ are $\cong$ because $\overline{BD}$ is bisector of $\overline{AC}$. $\triangle ABD$ and $\triangle CBD$ are rt. △s because $\perp$ form rt. $\angle$'s.

28. $\overline{AB} \cong \overline{DC}$ is needed because they are one pair of legs. $\overline{BC} \cong \overline{CB}$ by reflexive law and $\triangle ABC$ and $\triangle DCB$ are rt. △s because $\perp$ form rt. $\angle$'s.

29. *Proof:*

STATEMENTS	REASONS
1. $\overline{DB} \cong \overline{AC}$, $\overline{AB} \perp \overline{BC}$; $\overline{CD} \perp \overline{BC}$	1. Given
2. $\angle ABC$ and $\angle DCB$ are rt $\angle$'s	2. $\perp$ lines form rt. $\angle$'s
3. $\triangle ABC$ and $\triangle DCB$ are rt △s	3. Def. rt △s
4. $\overline{BC} \cong \overline{BC}$	4. Reflexive law
5. $\triangle ABC \cong \triangle DCB$	5. HL

Chapter 3 Practice Test

1. Exactly one by the Parallel Postulate.

2. True since $\ell \parallel m$, the 40° angle and $\angle 1$ are corresponding $\angle$'s so their measures are equal.

3. False, $m\angle 2 = 140°$ since it's supplementary to $\angle 1$.

4. True, since $\angle 1$ and $\angle 3$ are vertical angles, their measures are $=$.

5. False, the $\angle$'s are $\cong$ because they are corresponding $\angle$'s but not supplementary. The angles may appear to be right $\angle$'s but this is not known for sure since $\perp$ are not given.

6. The sum of the measures of the angles of an octagon is given by

$$S = (8-2)180° = (6)180° = 1080°.$$

7. 360° (the sum of the measures of the exterior angles of an polygon is 360°.)

8. The measure of each interior angle of a regular octagon is $\dfrac{1080°}{8} = 135°$ (1080° comes from Problem 6).

9. $m\angle 1 = 47°$; $\angle 1 \cong \angle 2$ because $a \parallel b$ so they are $\cong$ alt. int. $\angle$'s

$$6x + 5 = 9x - 16$$
$$-3x = -21$$
$$x = 7$$
$$m\angle 1 = 6(7) + 5 = 47°$$

10. $m\angle 3 = 144°$; Let $m\angle 1 = x$, then $m\angle 2 = x$ because they are $\cong$. Let $m\angle 3 = 4x$, $\angle 2$ and $\angle 3$ are supplementary.

$$m\angle 2 + m\angle 3 = 180$$
$$x + 4x = 180$$
$$5x = 180$$
$$x = 36$$
$$m\angle 3 = 4(36) = 144°$$

11. *Proof:*

STATEMENTS	REASONS
1. $\overline{AB} \cong \overline{CB}$, $\angle A \cong \angle C$ $\angle B$ is a right angle	1. Given
2. $\triangle ABE$ and $\triangle CBD$ are rt $\triangle\!\!\!\!{}_{\text{s}}$	2. Def. rt $\triangle$
3. $\triangle ABE \cong \triangle CBD$	3. LA

12. Let x = the measure of one congruent side in isos. $\triangle$

$$x + x + 12 = 46$$
$$2x = 34$$
$$x = 17$$

The other two sides of the triangle measure 17 cm each.

13. Corollary 3.16 states the measure of each angle of a regular polygon is $a = \dfrac{(n-2)180}{n}$ where n is the number of sides of the polygon.

$$162 = \dfrac{(n-2)180}{n} \quad \text{Multiply both sides of equation by } n$$
$$162n = (n-2)180$$
$$162n = 180n - 360$$
$$-18n = -360$$
$$n = 20$$

The polygon has 20 sides.

14. *Proof:*

STATEMENTS	REASONS
1. $\angle B \cong \angle D$; $\overline{AB} \parallel \overline{CD}$	1. Given
2. $\angle BAC \cong \angle DCA$	2. If lines are $\parallel$, alt. int. $\angle$'s $\cong$.
3. $\overline{AC} \cong \overline{AC}$	3. Reflexive law
4. $\triangle ABC \cong \triangle CDA$	4. AAS

15. 90°. Since the sum of the measures of $\angle$'s of a $\triangle = 180°$ and one angle measures 90°, the sum of the remaining two angles is 90°. $180° - 90° = 90°$

CHAPTERS 1-3 CUMULATIVE REVIEW

1. True   **2.** False

3. True   **4.** False

5. True   **6.** True   **7.** True

8. False   **9.** True

10. False   **11.** True

12. True   **13.** False

14. False   **15.** $63°$ ($90° - 27° = 63°$)

16.

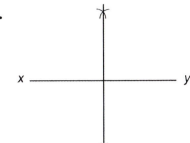

17. Scalene triangle   **18.** Obtuse triangle

19. $\angle A$ (or $\angle 1$)   **20.** $\angle 4$

21. $\overline{AB}$ **22.** 22.5 in. (7.5×3)

23.

STATEMENTS	REASONS
1. $\overline{AB} \cong \overline{BC}$	1. Given
2. $\angle A \cong \angle BCA$	2. If 2 sides of $\triangle \cong$, $\angle$'s opp. them are $\cong$.
3. $\overline{DE} \cong \overline{EC}$	3. Given
4. $\angle ECD \cong \angle D$	4. If 2 sides of $\triangle \cong$, $\angle$'s opp. them are $\cong$.
5. $\angle BCA \cong \angle ECD$	5. Vertical $\angle$'s $\cong$
6. $\angle A \cong \angle D$	6. Transitive law

24. $\overline{AD}$ is an altitude

$\overleftrightarrow{EH}$ is a perpendicular bisector

$\overline{CF}$ is a median

25.

STATEMENTS	REASONS
1. $\overline{WY}$ bisects $\angle XYZ$	1. Given
2. $\angle XYW \cong \angle ZYW$	2. Def. $\angle$ bisector
3. $\overline{YW}$ bisects $\angle XWZ$	3. Given
4. $\angle XWY \cong \angle ZWY$	4. Def. $\angle$ bisector
5. $\overline{WY} \cong \overline{WY}$	5. Reflexive law
6. $\triangle WXY \cong \triangle WZY$	6. AAS

26.   STATEMENTS   REASONS

1. $\overline{AD}$ and $\overline{BC}$ bisect each other   1. Given

2. $\overline{AE} \cong \overline{DE}$; $\overline{BE} \cong \overline{CE}$ 2. Def. of bisector

3. $\angle AEB \cong \angle DEC$ 3. Vertical $\angle$'s $\cong$

4. $\triangle AEB \cong \triangle DEC$ 4. SAS

5. $\overline{AB} \cong \overline{DC}$ 5. CPCTC

27. 54 mm; By Theorem 2.10, medians meet 2/3 distance from vertex to the midpoint of the opposite side.

$$AX = \frac{2}{3}(AE)$$

$$AX = \frac{2}{3}(162) = 108$$

$$XE = 162 - 108 = 54 \text{ mm}$$

28. **(a)** $29°$ **(b)** $92°$ **(c)** $122°$ **(d)** $58°$

(e) $122°$ **(f)** $58°$ **(g)** $97°$ **(h)** $54°$

(i) $59°$ **(j)** $63°$ **(k)** $23°$

29. $m\angle 1 = m\angle 4 = m\angle 5 = m\angle 8 = 133°$
$m\angle 2 = m\angle 3 = m\angle 6 = m\angle 7 = 47°$

$\angle 3$ and $\angle 6$ are alt. int $\angle$s. They are $\cong$ since $a \parallel b$.

$$6x + 5 = 9x - 16$$
$$3x = 21$$
$$x = 7$$

therefore $m\angle 3 = 6(7) + 5 = 47°$,
$m\angle 6 = 9(7) - 16 = 47$.

30. $1080°$; By Theorem 3.15 $S = (n-2)180°$ where n is number of sides of polygon $S = (8-2)180° = 1080°$.

31. $360°$ by Theorem 3.17

32.   STATEMENTS   REASONS

1. $\overline{AB} \perp \overline{BE}$; $\overline{DE} \perp \overline{BE}$; $\angle A \cong \angle D$ 1. Given

2. $\angle ABE$ and $\angle DEB$ are rt. $\angle$'s.   2. $\perp$ lines form rt $\angle$'s

3. $\triangle ABE$ and $\triangle DEB$ are rt. Ⓢ   3. Def. rt Ⓢ

4. $\overline{BE} \cong \overline{EB}$ 4. Reflexive law

5. $\triangle ABE \cong \triangle DEB$ 5. LA

33. $\triangle ABC \cong \triangle ETP$ by ASA

34. $\triangle FOR \cong \triangle QDE$ by SSS

35. $\triangle HOP \cong \triangle FIT$ by AAS

36. Not enough information

37. $\triangle ABC \cong \triangle EMC$ by ASA or LA

38. $\triangle WXZ \cong \triangle YXZ$ by AAS or HA

39. Not enough information

40. $\triangle SUM \cong \triangle FHG$ by SAS or LL

CHAPTER 4 QUADRILATERALS

Section 4.1 Parallelograms

4.1 PRACTICE EXERCISES

1. The reason for Statement 1 is: Given

Statement 2 is: $\overline{BD} \cong \overline{BD}$

Statement 3 is: $\triangle ABD \cong \triangle CBD$. Since $\overline{AB} \cong \overline{CD}$, $\overline{AD} \cong \overline{CB}$ and $\overline{BD} \cong \overline{BD}$ 3 sides of one triangle are congruent to the corresponding 3 sides of the other.

The reason for Statement 4 is: CPCTC
Note that $\angle 1$ and $\angle 4$ are corresponding angles in congruent triangles $\triangle ABD$ and $\triangle CDB$.
The reason for Statement 5 is: If alt. int. $\angle$'s are $\cong$ then lines are ‖.
Note that $\angle 1$ and $\angle 4$ are $\cong$ alternate interior angles formed by transversal $\overline{BD}$ making $\overline{AB}$ and $\overline{CD}$ parallel.

Statement 6 is: $\angle 2 \cong \angle 3$
This is similar to Statement 4.
Statement 7 is: $\overline{BC} \parallel \overline{AD}$
This is similar to Statement 5.

The reason for Statement 8 is: Def. of ▱
A quadrilateral with both parts of opposite sides parallel is a parallelogram.

2. The reason for Statement 1 is: Given

The reason for Statement 2 is: Def. of seg. bisector. This uses Statement 1.

Statement 3 is: $\angle 1 \cong \angle 2$
Since $\angle 1$ and $\angle 2$ are vertical angles, they are congruent.

Statement 4 is: $\triangle APB \cong \triangle CPD$. Since $\overline{AP} \cong \overline{CP}$ and $\overline{PD} \cong \overline{PB}$ included angles, $\angle 1$ and $\angle 2$ are $\cong$, the triangles are congruent.

The reason for Statement 5 is: CPCTC
$\angle 3$ and $\angle 4$ correspond in congruent triangles $\triangle APB$ and $\triangle CPD$.

The reason for Statement 6 is: Alt. int. $\angle$'s are $\cong$ if lines are ‖. Since $\angle 3$ and $\angle 4$ are alternate interior angles formed by transversal $\overline{BD}$, $\overline{AB}$ and $\overline{DC}$ are parallel.

The reason for Statement 7 is: CPCTC
$\overline{AB}$ and $\overline{DC}$ correspond in congruent triangles $\triangle APB$ and $\triangle CPD$.

Statement 8 is: $ABCD$ is a parallelogram
This follows since two sides $\overline{AB}$ and $\overline{DC}$ are $\cong$ and parallel opposite sides.

4.1 SECTION EXERCISES

1. True　　**3.** True　　**5.** True　　**7.** False

9. False　　**11.** True　　**13.** True　　**15.** True

17. False　　**19.** True

21. 1. Sum of $\angle$'s of quadrilateral $= 360°$
2. Given: Def. $\cong \angle$'s　3. Substitution law
5. Mult.-Div. Post.　6. $\angle A$ and $\angle B$ are supplementary.　7. $\overline{AD} \parallel \overline{BC}$　9. Def. supp. $\angle$'s.　10. $\overline{AB} \parallel \overline{DC}$　11. $ABCD$ is a parallelogram.

23. No, the four sides of two parallelograms could be congruent without the corresponding angles being congruent. [▱ ▱].

25.

STATEMENTS	REASONS
1. $\overline{VZ} \parallel \overline{WY}$; $\overline{VZ} \cong \overline{WX}$ $\overline{WY} \cong \overline{WX}$	1. Given
2. $\overline{VZ} \cong \overline{WY}$	2. Transitive law
3. $VWYZ$ is ▱	3. If one pair of opp. sides of quad. are $\cong$ and $\parallel$, the quad is ▱.

27. If one angle of a parallelogram is twice another, they must be consecutive angles. Suppose one angle is x, then the consecutive angle is $2x$. Since the angles opposite each of these are also x and $2x$, using the fact that the angles of a parallelogram sum to $360°$, we must solve:

$$x + x + 2x + 2x = 360$$
$$6x = 360$$
$$x = 60$$
$$2x = 120$$

Thus, the angles measure $60°$ and $120°$. An alternative solution would use the fact that consecutive $\angle$'s of ▱ are supplementary.

29. *Proof:*

STATEMENTS	REASONS
1. $\overline{AP} \cong \overline{QC}$	1. Given
2. $ABCD$ is a parallelogram	2. Given
3. $\angle 1 \cong \angle 2$	3. If lines $\parallel$, alt. int. $\angle$'s are $\cong$
4. $\overline{AB} \cong \overline{DC}$	4. Opp. sides of ▱ are $\cong$
5. $\triangle ABP \cong \triangle CDQ$	5. SAS
6. $\overline{PB} \cong \overline{QD}$	6. CPCTC
7. $\angle 3 \cong \angle 4$	7. If lines $\parallel$, alt. int. $\angle$'s are $\cong$
8. $\overline{AD} \cong \overline{CB}$	8. Opp. sides of ▱ are $\cong$
9. $\triangle APD \cong \triangle CQB$	9. SAS
10. $\overline{PD} \cong \overline{QB}$	10. CPCTC
11. $PBQD$ is a parallelogram	11. If opp. sides are $\cong$, then ▱

31. ▱

33.

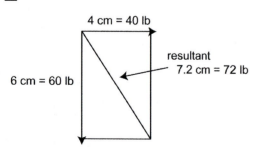

approximately 72 pounds

Actually draw this parallelogram with one side 4 cm and the adjacent side 6 cm. Measure the resultant vector and multiply by a factor of 10 since a length of 4 cm = 40 pounds.

Section 4.2 Rhombus and Kite

4.2 PRACTICE EXERCISES

1. Statement 1 is: $\overline{AC} \perp \overline{BD}$

The reason for Statement 2 is: Def. of $\perp$
Recall that $\perp$ lines form congruent adjacent angles

The reason for Statement 3 is: Reflexive law

The reason for Statement 4 is: Given
Remember that $\square$ $ABCD$ is the notation to indicate that $ABCD$ is a parallelogram.

Statement 5 is: $\overline{AP} \cong \overline{CP}$
Since the diagonals of a $\square$ bisect each other, $\overline{BD}$ bisects $\overline{AC}$ making $\overline{AP} \cong \overline{CP}$.

Statement 6 is: $\triangle ABP \cong \triangle CBP$
Since $\overline{BP} \cong \overline{BP}$, $\overline{AP} \cong \overline{CP}$ and $\angle 1 \cong \angle 2$, the triangles are congruent by SAS.

The reason for Statement 7 is: CPCTC

Statement 8 is: $\square$ $ABCD$ is rhombus
A parallelogram with two equal adjacent sides ($\overline{AB}$ and $\overline{CB}$) is a rhombus.

2. Reason 2: Def. of rhombus. A rhombus is a $\square$ with $2 \cong$ adj. sides

Reason 3: Reflexive law

Reason 4: Diagonals of $\square$ bisect each other. Remember a rhombus is $\square$ thus it has all the properties of $\square$.

Reason 5: SSS

Reason 6: CPCTC

Reason 7: Def. $\angle$ bisector.

3. Reason 1: Given

Reason 2: Def. of kite. Recall a kite is quadrilateral with exactly 2 distinct pairs of $\cong$ consecutive sides.

Statement 3: $\overline{BD} \cong \overline{BD}$ Reason 3: Reflexive law

Statement 4: $\triangle ABD \cong \triangle CBD$. Statements 2 and 3 show the 3 pairs of corresponding sides $\cong$.

Statement 5: $\angle 1 \cong \angle 2$; $\angle 3 \cong \angle 4$. This is needed to show $\overline{BD}$ bisects $\angle ABC$ and $\angle ADC$.

Reason 6: Def. of $\angle$ bisector. The angle bisector is a line segment that separates the angle into $2 \cong$ adj. $\angle$'s.

4.2 SECTION EXERCISES

1. (a) Isosceles $\triangle$ because $\triangle ABC$ and $2 \cong$ sides since $ABCD$ is a rhombus which has $4 \cong$ sides.

(b) Right $\triangle$ because the diagonals of a rhombus are $\perp$.

(c) Yes. One reason is SSS. $\overline{AD} \cong \overline{AB}$ because they are sides of a rhombus.
$\overline{BE} \cong \overline{DE}$ because the diagonals of a rhombus bisect each other. Finally $\overline{AE} \cong \overline{AE}$ by the Reflexive law.

3. A rhombus and a kite are similar in that they are both quadrilaterals, diagonals are $\perp$, have $2 \cong$ adj. sides and one diagonal bisects the other diagonal.
The differences are a rhombus is a parallelogram, so it has all the properties of a parallelogram. A kite is not a parallelogram.

5. 50.4 in (4(12.6))

7. 4.2 in (2(1.6)+2(0.5))

9. $m\angle 1 = 60$, $m\angle 2 = 60°$ The marks indicate the diagonal of the rhombus is the same length as each side. Think of this rhombus as 2 equilateral $\triangle_S$ with a common side. Thus $\angle 1$ and $\angle 2$ must each measure 60°.

11. $m\angle 2 = 121°$ since opposite $\angle$'s of a rhombus are $\cong$. The sum of $\angle$'s of quadrilateral = 360°. The sum of the remaining two $\cong$ angles is $360 - (121 + 121) = 118$. One angle measures $\dfrac{118}{2} = 59°$.

$m\angle 1 = 59°$

13. *Proof:*

STATEMENTS	REASONS
1. $\square$ *ABCD* is a rhombus	1. Given
2. $\overline{AC} \perp \overline{BD}$	2. Diag. of rhombus are $\perp$
3. $m\angle DPC = 90°$	3. $\perp$ lines form 90° angles
4. $m\angle DPC + m\angle 1 + m\angle 2 = 180°$	4. Sum of $\angle$'s of $\triangle = 180°$
5. $90° + m\angle 1 + m\angle 2 = 180°$	5. Substitution
6. $m\angle 1 + m\angle 2 = 90°$	6. Add.-Subt. Post.
7. $\angle 1$ and $\angle 2$ are complementary	7. Def. of comp. $\angle$'s

---

Section 4.3 Rectangles and Squares

4.3 PRACTICE EXERCISES

1. 1. True because a square is a rectangle and a rectangle is a $\square$.

2. False, because a rhombus does not have to have all right $\angle$'s.

3. True because all the angles are right $\angle$'s.

4. True by the definition of a square.

5. False, only the diagonals of a rectangle, rhombus or square are $\cong$.

2. Since $\overline{PR}$ is the base of isosceles $\triangle PQR$, the legs are $\overline{RQ}$ and $\overline{PQ}$. With $RQ = 30$ inches, $PQ = 30$ inches also (the legs are equal). Since M and O are midpoints of $\overline{PR}$ and $\overline{QR}$, respectively, $\overline{MO}$ is parallel to $\overline{PQ}$ and equal to one-half of PQ by Theorem 4.19. Thus,

$$MQ = \frac{1}{2}PQ = \frac{1}{2}(30) = 15 \text{ inches.}$$

4.3 SECTION EXERCISES

		Parallelogram ▱	Rhombus ◇	Kite ◇	Rectangle ▭	Square ◻
	Definitions → Properties ↓	Quad. in which both pairs of opposite sides are parallel	Parallelogram with two congruent adjacent sides.	Quad. with exactly two distinct pairs of ≅ adjacent sides.	Parallogram that has a right angle.	Rhombus with a right angle.
1.	Both pairs of Opposite sides ‖	✓	✓		✓	✓
2.	Diagonals form 2 ≅ △	✓	✓		✓	✓
3.	Opposite ∠'s ≅	✓	✓		✓	✓
4.	Opposite sides ≅	✓	✓		✓	✓
5.	Diagonals bisect each other	✓	✓		✓	✓
6.	Consecutive ∠'s supplementary	✓	✓		✓	✓
7.	All ∠'s are right ∠'s				✓	✓
8.	Diagonals are ≅				✓	✓
9.	Diagonals are ⊥		✓	✓		✓
10.	All sides ≅		✓			✓
11.	2 ≅ adjacent sides		✓	✓		✓
12.	One diagonal ⊥ bisector of other		✓	✓		✓
13.	Diagonals bisect ∠'s		✓			✓

15. 8   **17.** 8

19. 60 inches (10 + 24 + 26) because opposite sides of a rectangle are ≅ and diagonals of a rectangle bisect each other.

21. 50 inches (13 + 13 + 24) because opposites sides of a rectangle are ≅ and diagonals of a rectangle bisect each other.

23. Use Construction 4.2

25. Since $DE = \frac{1}{2}AB$ ($\overline{DE}$ joins the midpoints of two sides of $\triangle ABC$ making it half the third side), $56 = \frac{1}{2}AB$; so $AB = 112$ ft.

27. (a) $XY = \dfrac{1}{2}(AC) = \dfrac{1}{2}(18) = 9$ cm

(b) $YZ = \dfrac{1}{2}(AB) = \dfrac{1}{2}(10) = 5$ cm

(c) $XZ = \dfrac{1}{2}(BC) = \dfrac{1}{2}(14) = 7$ cm

(d) $10 + 14 + 18 = 42$ cm

(e) $9 + 5 + 7 = 21$ cm

(f) The perimeter of the $\triangle$ formed by joining the midpoints is one half the perimeter of $\triangle ABC$. Yes the perimeter will always be one half the perimeter of the original $\triangle$ because the lengths of the sides of the first $\triangle$ are always ½ the lengths of the original sides by Theorem 4.19.

29. False,

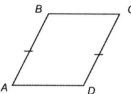

31. True, because all the $\angle$'s of a rectangle are rt $\angle$'s.

33. 204 squares
There is 1, 8×8 square on the board
There are 4, 7×7 squares on the board
There are 9, 6×6 squares on the board.
Notice the pattern: $1^2, 2^2, 3^2$ etc.
The total number of squares on the board is

$$1^2 + 2^2 + 3^2 + 4^2 + 5^2 + 6^2 + 7^2 + 8^2$$
$$= 64 + 49 + 36 + 25 + 16 + 9 + 4 + 1 = 204$$

35. $m\angle 1 = 59°$ $\quad m\angle 2 = 121°$ $\quad m\angle 3 = 121°$
$m\angle 4 = 31°$ $\quad m\angle 5 = 39°$ $\quad m\angle 6 = 20°$
$m\angle 7 = 90°$ $\quad m\angle 8 = 20°$ $\quad m\angle 9 = 62°$
$m\angle 10 = 28°$ $\quad m\angle 11 = 90°$ $\quad m\angle 12 = 90°$
$m\angle 13 = 31°$ $\quad m\angle 14 = 59°$ $\quad m\angle 15 = 62°$
$m\angle 16 = 59°$ $\quad m\angle 17 = 59°$ $\quad m\angle 18 = 121°$
$m\angle 19 = 62°$ $\quad m\angle 20 = 28°$

Section 4.4 Trapezoids

1. True **3.** True **5.** True **7.** True

9. True **11.** False **13.** True **15.** True

17. True **19.** False

21. 4.5 cm by Theorem 4.23

23. 81° by Theorem 3.8

25. 63° by Theorem 3.8

27. 99° by Theorem 3.11

29. $\dfrac{4.8 + 13.2}{2} = 9$

31. $\dfrac{(2x+5)+(6x-1)}{2} = AB$

$\qquad \dfrac{8x+4}{2} = AB$

$\qquad 4x+2 = AB$

33. 1. By Construction   2. $ABCD$ is a trapezoid with median $\overline{EF}$ 3. Def. of median of trapezoid   4. $\overline{AE} \cong \overline{ED}$ and $\overline{BF} \cong \overline{FC}$

5. Vert. $\angle$'s are $\cong$ 6. $\overline{AB} \parallel \overline{CD}$ 7. If lines are $\parallel$, alt. int. $\angle$'s are $\cong$ 8. AAS
9. CPCTC   11. Seg.-Add. Post.
12. Substitution law
15. Line $\parallel$ to one of two $\parallel$ lines is $\parallel$ to other.

35. *Given:* Isosceles trapezoid *ABCD* with equal

sides $\overline{AD}$ and $\overline{BC}$ and diagonals

$\overline{AC}$ and $\overline{BD}$

Prove: $\overline{AC} \cong \overline{BD}$

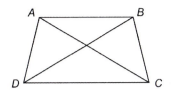

Proof: STATEMENTS	REASONS
1. *ABCD* is an isosceles trapezoid with $\overline{AD} \cong \overline{BC}$	1. Given
2. $\angle ADC \cong \angle BCD$	2. Base $\angle$'s of isosc. trapezoid are $\cong$
3. $\overline{DC} \cong \overline{DC}$	3. Reflexive law
4. $\triangle ADC \cong \triangle BCD$	4. SAS
5. $\overline{AC} \cong \overline{BD}$	5. CPCTC

37. *Given:* ▱ *ABCD* with *P, Q, R,* and *S*
midpoints of adjacent sides

Prove: *PQRS* is a rhombus

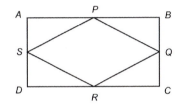

Proof: STATEMENTS	REASONS
1. *ABCD* is a rectangle with *P, Q, R,* and *S* midpoints of sides	1. Given
2. $\angle A$, $\angle B$, $\angle C$, and $\angle D$ are right angles	2. All $\angle$'s of ▭ are right angles
3. $\angle A \cong \angle B \cong \angle C \cong D$	3. All rt. $\angle$'s are $\cong$
4. $\overline{AB} \cong \overline{DC}$ and $\overline{AD} \cong \overline{BC}$	4. Opp. sides of ▭ are $\cong$
5. $\overline{AP} \cong \overline{PB}$, $\overline{BQ} \cong \overline{QC}$, $\overline{CR} \cong \overline{DR}$, and $\overline{DS} \cong \overline{AS}$	5. Def. of midpoint
6. $\triangle APS \cong \triangle BPQ \cong \triangle CRQ \cong \triangle DRS$	6. SAS
7. $\overline{SP} \cong \overline{QP} \cong \overline{QR} \cong \overline{SR}$	7. CPCTC
8. *PQRS* is a parallelogram	8. Both pairs of opp. sides are $\cong$
9. *PQRS* is a rhombus	9. Two adj. sides are $\cong$ in a ▱

Chapter 4 Review Exercises

1. True   **2.** True   **3.** False   **4.** True   **5.** True   **6.** False   **7.** False   **8.** True

9. True   **10.** True

11. *Proof:*   STATEMENTS   REASONS

       1. $\angle 1 \cong \angle 2$ 1. Given

       2. $\overline{NO} \cong \overline{NP}$ 2. If 2$\angle$'s in a $\triangle \cong$, sides opp. them are $\cong$.

       3. $\overline{MQ} \cong \overline{NO}$, $\overline{MN} = \overline{QP}$ 3. Given

       4. $\overline{MQ} \cong \overline{NP}$ 4. Transitive law

       5. *MNPQ* is $\square$ 5. If opp. sides of quad. $\cong$, it's a $\square$.

12. *Proof:*   STATEMENTS   REASONS

       1. $\overline{AB} \cong \overline{CD}$; $\angle 1 \cong \angle 2$ 1. Given

       2. $\overline{AB} \parallel \overline{CD}$ 2. If alt. int. $\angle$'s $\cong$, lines $\parallel$.

       3. *ABCD* is parallelogram   3. If 2 opp. sides of quad. $\cong$ and $\parallel$, then $\square$.

13. *Proof*   STATEMENTS   REASONS

       1. $\angle 1 \cong \angle 2$; $\angle 3 \cong \angle 4$ 1. Given

       2. $\overline{NL} \cong \overline{NL}$ 2. Reflexive law

       3. $\triangle MNL \cong \triangle ONL$ 3. ASA

       4. $\overline{MN} \cong \overline{ON}$ and $\overline{ML} \cong \overline{OL}$ 4. CPCTC

       5. *LMNO* is kite   5. Def. kite (2 pr. $\cong$ consec. sides)

14. *Proof:*   STATEMENTS   REASONS

       1. $\square$ *ABCD*, $\square$ *DEFG*   1. Given

       2. $\angle B \cong \angle CDA$ 2. Opp. $\angle$'s $\square$ $\cong$

       3. $\angle CDA \cong \angle EDG$ 3. Vertical $\angle$'s $\cong$

       4. $\angle EDG \cong \angle F$ 4. Opp. $\angle$'s $\square$ $\cong$

       5. $\angle B \cong \angle F$ 5. Transitive law

15. $m\angle 1 = 108°$ by Theorem 4.13, one pair of opp. $\angle$'s $\cong$ in a kite. The sum of the $\angle$'s of quadrilateral is $360°$ by Theorem 3.15. $m\angle 2 = 360 - (108 + 108 + 43) = 101°$

16. *Proof:*   STATEMENTS   REASONS

1. $\triangle EBF \cong \triangle GDH$ 1. Given
    $\triangle AEH \cong \triangle CGF$
2. $\overline{EF} \cong \overline{GH}$; $\overline{EH} \cong \overline{GF}$ 2. CPCTC

3. $EFGH$ ▱   3. If both pr. opp. sides of quad $\cong$, then ▱.

17. False   **18.** True   **19.** True   **20.** True   **21.** True   **22.** False   **23.** False

24. 50.8 cm   **25.** True   **26.** False   **27.** True   **28.** False   **29.** True   **30.** True

31. False   **32.** True

33. $AD = BC$
    $6x - 1 = 5x + 2$
    $x = 3$
    $CD = AB = 2(3) + 3 = 9$

34. $DE = \dfrac{1}{2}(BC)$
    $9x - 4 = \dfrac{1}{2}(6x + 4)$
    $18x - 8 = 6x + 4$
    $12x = 12$
    $x = 1$
    $DE = 5$; $BC = 10$

35. *Proof:*   STATEMENTS   REASONS

1. Rectangle $ABCD$ with diagonals $\overline{AC}$ and $\overline{BD}$ 1. Given

2. $\overline{AB} \cong \overline{DC}$ 2. Opp. sides rect. $\cong$

3. $\overline{BD} \cong \overline{CA}$ 3. Diagonals of rect. $\cong$

4. $\overline{AD} \cong \overline{AD}$ 4. Reflexive law

5. $\triangle ABD \cong \triangle DCA$ 5. SSS

6. $\angle 1 \cong \angle 2$ 6. CPCTC

36. Use Construction 1.3 to construct the two diagonals, they are $\perp$ bisectors of each other. Draw the square by connecting endpoints of diagonals.

37. *Proof:*   STATEMENTS   REASONS

1. $ABCD$ is a rectangle   1. Given

2. $BCDE$ is a parallelogram   2. Given

3. $\overline{AC} \cong \overline{BE}$ 3. Diag. of rect. are $\cong$

4. $\overline{BE} \cong \overline{CD}$ 4. Opp. sides of ▱ are $\cong$.

5. $\overline{AC} \cong \overline{CD}$ 5. Transitive law

6. $\triangle ACD$ is isosceles   6. Def. of isosc. $\triangle$

38. False   **39.** True   **40.** True

41. False   **42.** True

43. No, example: 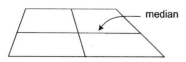 ← median

44. Use Const. 4.3

45. $\dfrac{32+40}{2} = 36$ ft

46. $\dfrac{BC+AD}{2} = xy$

$\dfrac{3+(AD)}{2} = 12.6$

$3+(AD) = 25.2$

$AD = 22.2$

47. $m\angle D = 67°$ because the base $\angle$'s isos. trapezoid $\cong$. $\angle B \cong \angle C$ for the same reason. $m\angle B + m\angle B + 67° + 67° = 360°$ by Theorem 3.15

$m\angle B = 113° = m\angle C$

Thus $m\angle B = 113°$, $m\angle C = 113°$, $m\angle D = 67°$

Chapter 4 Practice Test

1. True in a kite, square and rhombus but not rectangle or other $\square$.

2. True by the definition of a square.

3. False, only true in an isosceles trapezoid.

4. True because a rhombus is $\square$ and opp. $\angle$'s are $\cong$ in all $\square$.

5. $m\angle ADC = 70°$, since $m\angle BCD = 110°$ and consecutive $\angle$'s of $\square$ are supp. $m\angle ADC = 180° - 110° = 70°$.

6. *Proof:*

STATEMENTS	REASONS
1. $ABCD$ is rectangle, $\overline{AE} \cong \overline{FC}$	1. Given
2. $\angle A$ and $\angle C$ are rt. $\angle$'s	2. All $\angle$'s rect. are rt. $\angle$'s
3. $\triangle ADE$ and $\triangle CBF$ are rt. $\triangle$.	3. Def. rt $\triangle$
4. $\overline{AD} \cong \overline{CB}$	4. Opp. sides rect. $\cong$
5. $\triangle ADE \cong \triangle CBF$	5. LL
6. $\overline{DE} \cong \overline{BF}$	6. CPCTC

7. *Given: ABCD* is a rectangle with $\overline{AE} \cong \overline{FC}$

Prove: DEBF is a parallelogram

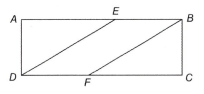

Proof: STATEMENTS   REASONS

1. *ABCD* is a rectangle with $\overline{AE} \cong \overline{FC}$ 1. Given; def. $\cong$ seg.
   therefore *AE = FC*

2. $\overline{EB} \parallel \overline{DF}$ 2. $\overline{AB} \parallel \overline{DC}$ since they are opp. sides
                                                                            of a $\square$.

3. $\overline{AB} \cong \overline{DC}$ therefore *AB = DC*   3. Opp. sides of $\square$ are $\cong$; def. $\cong$ seg.

4. *AB = AE + EB* and *DC = DF + FC*   4. Seg.-Add. Post.

5. *AE + EB = DF + FC*   5. Substitution law (Steps 3 and 4)

6. *AE + EB = DF + AE*   6. Substitution (Steps 1 and 5)

7. *EB = DF*   7. Add.-Subt. Post.

8. *DEBF* is a parallelogram   8. Opp. sides are $\cong$ and $\parallel$

8. Use Construction 1.2 and a construction similar to 4.2.

9. 88 cm because the length of a segment joining the midpoints of 2 sides $\triangle$= one half length of third side.

10. $MN = \frac{1}{2}(AC)$

$x + 9 = \frac{1}{2}(4x + 6)$

$x + 9 = 2x + 3$

$6 = x$

$AC = 30$

11. Perimeter $= 2(AD) + 2(DC)$ where $AB = AD$ and $BC = DC$ by def. Kite.
Let $x = DC$

$54.6 = 2(12.4) + 2x$

$54.6 = 24.8 + 2x$

$29.8 = 2x$

$14.9 = x$

$DC = 14.9$ in

12. Since $\overline{AD} \parallel \overline{BC}$, $\angle B$ and $\angle A$ are supplementary, therefore $m\angle B = 113$. Because the legs of the trapezoid are $\cong$, it's an isosceles trapezoid where base $\angle$'s are $\cong$. $m\angle B = m\angle C = 113°$

13. $MN = \frac{1}{2}(WX + YZ)$

$MN = \frac{1}{2}(17 + 31)$

$MN = 24$ cm

14. $MN = \frac{1}{2}(WX + YZ)$

$2x + 11 = \frac{1}{2}(4x - 7 + 2x + 1)$

$2x + 11 = \frac{1}{2}(6x - 6)$

$2x + 11 = 3x - 3$

$14 = x$

15. Yes, the two quadrilaterals with ≅ diagonals are a square and a rectangle. Since a square is a rectangle by definition, the figure must be a rectangle.

16. *Proof:*   STATEMENTS   REASONS

      1. $ABCD$ ▱ with $\overline{AD} \cong \overline{CN}$, $\angle 1 \cong \angle 2$ 1. Given

      2. $\overline{CN} \cong \overline{CD}$ 2. If 2∠'s△ are ≅, sides opp. those ∠'s ≅

      3. $\overline{AD} \cong \overline{CD}$ 3. Transitive law

      4. $ABCD$ is rhombus   4. Def. rhombus (▱ with 2 ≅ adj. sides)

17. Square

18. Square or rhombus

19. Trapezoid

20. Kite, square or rhombus

CHAPTER 5   SIMILAR POLYGONS AND THE PYTHAGOREAN THEOREM

Section 5.1 Ratio and Proportion

5.1 PRACTICE EXERCISES

1. $\dfrac{14}{10} = \dfrac{7}{5}$

2. Yes, $\dfrac{1}{5} = \dfrac{2}{10} = \dfrac{3}{15} = \dfrac{4}{10}$

3. The reason for Statement 1 is: Given

   The reason for Statement 2 is: Mult.-Div. Prop.

   For example, since $x = \dfrac{a}{b}$, multiplying both sides by b gives $bx = a$ or $a = bx$.

   The reason for Statement 3 is: Add.-Subt. Prop.

Statement 4 is: $a + c + e = (b + d + f)x$

This follows directly from the distributive law

The reason for Statement 5 is: Mult.-Div. Prop.

This follows since $a + c + e = (b + d + f)x$ so we divide both sides by $b + d + f$ to obtain the result.

Statement 6 is: $\dfrac{a + c + e}{b + d + f} = \dfrac{a}{b}$

This follows since we have both fractions equal to x hence to each other.

5.1 SECTION EXERCISES

1. $\dfrac{20}{35} = \dfrac{4 \cdot \cancel{5}}{7 \cdot \cancel{5}} = \dfrac{4}{7}$

3. $\dfrac{8 \text{ cm}}{32 \text{ cm}} = \dfrac{\cancel{8} \cdot 1 \ \cancel{cm}}{4 \cdot \cancel{8} \ \cancel{cm}} = \dfrac{1}{4}$

5. $\dfrac{200 \text{ mi}}{4 \text{ hr}} = \dfrac{\cancel{4} \cdot 50}{\cancel{4}} \dfrac{\text{mi}}{\text{hr}} = 50 \dfrac{\text{mi}}{\text{hr}} = 50 \text{ mph}$

7. $\dfrac{4 \text{ in}}{2 \text{ ft}} = \dfrac{4 \text{ in}}{2(12) \text{ in}} = \dfrac{\cancel{4} \ \cancel{in}}{\cancel{4} \cdot 6 \ \cancel{in}} = \dfrac{1}{6}$

9. $\dfrac{\frac{1}{2} \ \cancel{in}}{\frac{3}{8} \ \cancel{in}} = \dfrac{1}{2} \div \dfrac{3}{8} = \dfrac{1}{2} \cdot \dfrac{8}{3} = \dfrac{8}{2 \cdot 3} = \dfrac{\cancel{2} \cdot 4}{\cancel{2} \cdot 3} = \dfrac{4}{3}$

11. $\dfrac{a}{3} = \dfrac{14}{21}$
    $21a = (14)(3)$ Means-extremes property
    $21a = 42$
    $a = \dfrac{42}{21} = 2$

13. $\dfrac{40}{35} = \dfrac{2}{y}$
    $40y = (2)(35)$ Means-extremes property
    $40y = 70$
    $y = \dfrac{70}{40} = \dfrac{7}{4}$

15. $\dfrac{25}{x} = \dfrac{x}{1}$
    $(25)(1) = (x)(x)$ Means-extremes property
    $25 = x^2$

    Since $5^2 = 25$ and $(-5)^2 = 25$, $x = 5$ or $x = -5$.

17.
$$\frac{a+2}{a} = \frac{7}{5}$$

$5(a+2) = 7a$ Means-extremes property

$5a + 10 = 7a$ Distributive law

$10 = 2a$ Subtract $5a$ from both sides

$5 = a$

19.
$$\frac{y+2}{12} = \frac{y-2}{4}$$

$4(y+2) = 12(y-2)$ Means-extremes property

$4y + 8 = 12y - 24$ Distributive law

$8 = 8y - 24$ Subtract $4y$ from both sides

$32 = 8y$ Add 24 to both sides

$4 = y$

21. Let x be the geometric mean between 4 and 36. Then we must solve:

$$\frac{4}{x} = \frac{x}{36}$$

$(4)(36) = (x)(x)$ Means-extremes property

$144 = x^2$

Since $(12)^2 = 144$ and $(-12)^2 = 144$, $x = 12$ or $x = -12$.

The geometric mean is 12 or −12.

23. Let x be the geometric mean between 9 and 64. Then we must solve:

$$\frac{9}{x} = \frac{x}{64}$$

$(9)(64) = (x)(x)$ Means-extremes property

$576 = x^2$

Since $(24)^2 = 576$ and $(-24)^2 = 576$, $x = 24$ or $x = -24$.
The geometric mean is 24 or −24.

25. Let x = No. votes for Wettaw's opponent. The proportion that describes the information given is:

$$\frac{8}{5} = \frac{10,400}{x}$$

$8x = (10,400)(5)$ Means-extremes property

$8x = 52,000$

$x = \dfrac{52,000}{8} = 6500$

Thus, Representative Wettaw's opponent received 6500 votes.

27. Let x = No. miles represented by $6\frac{1}{2}$ inches.

The proportion that describes the information given is:

$$\frac{\frac{1}{2}}{10} = \frac{6\frac{1}{2}}{x}$$

$\left(\dfrac{1}{2}\right)(x) = \left(6\dfrac{1}{2}\right)(10)$ Means-extremes property

$\dfrac{1}{2}x = \left(\dfrac{13}{2}\right)(10)$

$2\left(\dfrac{1}{2}x\right) = 2\left(\dfrac{13}{2}\right)(10)$

$\phantom{2\left(\dfrac{1}{2}\right)}x = (13)(10) = 130$

Thus, $6\frac{1}{2}$ inches on the map represents 130 mi.

29. Let x = No. pounds of garbage produced by 7 families. The proportion that describes the information given is:

$$\frac{1}{115} = \frac{7}{x}$$

Notice that in "The family of four", the "four" is not used in the proportion.

$(1)(x) = (115)(7)$ Means-extremes property

$x = 805$

Thus, 7 families will produce 805 lb of garbage in one week.

31. Take the appropriate measurements and approximate the various ratios.

33. Represent the lengths of the sides of the $\triangle$ as $3x, 4x$ and $5x$.

$$3x + 4x + 5x = 90$$
$$12x = 90$$
$$x = 7.5$$

The sides measure 22.5 cm, 30 cm and 37.5 cm.

35. Represent the measures of the angles as $2x$ and $7x$.

$$2x + 7x = 180$$
$$9x = 180$$
$$x = 20$$

The angles measure $40°$ and $140°$.

37. A ratio compares numbers by using division. A ratio is written as a fraction. A proportion is an equation that shows two ratios are equal.

Section 5.2 Similar Polygons

5.2 PRACTICE EXERCISES

1. In the figure $ABCD \sim A'B'C'D$. Since $\overline{D'C'}$ corresponds to $\overline{DC}$ and $DC = 15$ and D'C' = y, and also $\overline{A'D'}$ corresponds to $\overline{AD}$ with $A'D' = 5$ and $AD = 10$, we have

$$\frac{A'D'}{AD} = \frac{D'C'}{DC}.$$

Substituting we have:

$$\frac{5}{10} = \frac{y}{15}$$
$$(5)(15) = 10y \quad \text{Means-extreme prop.}$$
$$75 = 10y \quad \text{Simplify}$$
$$\frac{75}{10} = y \quad \text{Mult.-Div. Prop.}$$
$$7.5 = y$$

Since $\angle D'$ corresponds to $\angle D$ and $m\angle D = 129°$, $m\angle D' = 129°$.

2. Since $\angle ACB \cong \angle DCE$ (they are vertical angles), and $\angle B \cong \angle D$ (they are right angles), $\triangle ABC \sim \triangle EDC$

$$\frac{AC}{CE} = \frac{AB}{DE}.$$

Substituting we have:

$$\frac{AC}{10} = \frac{9}{6}$$
$$6AC = (10)(9) \quad \text{Means-extremes prop.}$$
$$6AC = 90$$
$$AC = \frac{90}{6} = 15 \text{ ft}$$

Also, $\dfrac{BC}{CD} = \dfrac{AB}{DE}.$

Substituting we have:

$$\frac{12}{CD} = \frac{9}{6}$$
$$(12)(6) = 9CD \quad \text{Means-extremes prop.}$$
$$72 = 9CD$$
$$\frac{72}{9} = CD$$
$$8 = CD$$

Thus, $CD = 8$ ft.

3. The reason for Statement 1 is: Given

Statement 2 is: $\angle 1 \cong 2$

The reason Statement 3 is: By construction a line can be constructed through a point not on a line parallel to that line.

Statement 4 is: $\angle 3 \cong \angle 1$
Note that $\angle 3$ and $\angle 1$ are corresponding angles formed when transversal $\overline{EC}$ cuts parallel lines $\overline{BE}$ and $\overline{AD}$.

The reason for Statement 5 is: If lines ∥, alt. int. $\angle$'s are $\cong$. Note that $\angle 2$ and $\angle 4$ are alternate interior angles formed when transversal $\overline{AB}$ cuts parallel lines $\overline{BE}$ and $\overline{AD}$.

The reason for Statement 6 is: Trans. laws
Statement 7 is: $\overline{AE} \cong \overline{AB}$ so $AE = AB$
Sides $\overline{AE}$ and $\overline{AB}$ of $\triangle ABE$ are opposite $\cong$ angles $\angle 3$ and $\angle 4$.

The reason for Statement 8 is: $\triangle$ Proportionality Theorem.
Since $\overline{AD}$ is ∥ to $\overline{BE}$, it divides the other two sides $\overline{EC}$ and $\overline{BC}$ into proportional segments.

The reason for Statement 9 is: Substitution law.
Since $AE = AB$, substitute into the formula in Statement 8.

5.2 SECTION EXERCISES

1. $\dfrac{x}{5} = \dfrac{12}{4}$

$4x = 5 \cdot 12$

$x = \dfrac{5 \cdot 12}{4} = 5 \cdot 3 = 15$ cm

3. $\dfrac{z}{4} = \dfrac{12}{4}$

$4z = 4 \cdot 12$

$z = \dfrac{4 \cdot 12}{4} = 12$ cm

5. The sum of the angles of a hexagon is

$$S = (6-2)180° = (4)(180°) = 720°.$$

Thus,

$$m\angle A' + m\angle B' + m\angle C' + m\angle D' + m\angle E' + m\angle F' = 720°$$
$$115° + 147° + \angle C' + 115° + 128° + 112° = 720°$$
$$m\angle C' + 617° = 720°$$
$$m\angle C' = 103°.$$

7. Always

9. Sometimes

11. Always

13. Sometimes

15. Never

17. Sometimes

19. By Theorem 5.11

$$\frac{AD}{DB} = \frac{AE}{EC}.$$

Substitute 12 for AD, 6 for DB, and 20 for AE.

$$\frac{12}{6} = \frac{20}{EC}$$
$$12EC = 6 \cdot 20$$
$$EC = \frac{6 \cdot 20}{12} = \frac{20}{2} = 10 \text{ ft}$$

21. Since $AB = AD + BD$ and $AB = 22$, $22 = AD + BD$. Also $AC = AE + EC = 8 + 3 = 11$. Since $\triangle ADE \sim \triangle ABC$

$$\frac{AD}{AB} = \frac{AE}{AC}.$$

Substitute 22 for AB, 8 for AE, and 11 for AC.

$$\frac{AD}{22} = \frac{8}{11}$$
$$11AD = 8 \cdot 22$$
$$AD = \frac{8 \cdot 22}{11} = 8 \cdot 2 = 16$$

Then $22 = AD + BD$ so $22 = 16 + BD$ giving $BD = 22 - 16 = 6$. Thus, $AD = 16$ yd and $BD = 6$ yd.

23. Since $\triangle ADE \sim \triangle ABC$,

$$\frac{AD}{AB} = \frac{AE}{AC}.$$

Substitute 18 for AC, 12 for AE, and 15 for AB.

$$\frac{AD}{15} = \frac{12}{18}$$
$$18AD = 12 \cdot 15$$
$$AD = \frac{12 \cdot 15}{18} = \frac{2 \cdot 15}{3} = 2 \cdot 5 = 10$$

Since $AB = AD + BD$, $15 = 10 + BD$ making $BD = 15 - 10 = 5$. Thus, $AD = 10$ ft and $BD = 5$ ft.

25. By Theorem 5.12

$$\frac{BD}{DC} = \frac{AB}{AC}.$$

Substitute 6 for AB, 9 for AC, and 2 for BD.

$$\frac{2}{DC} = \frac{6}{9}$$
$$2 \cdot 9 = 6DC$$
$$\frac{2 \cdot 9}{6} = DC$$
$$3 = DC$$

Thus $DC = 3$ ft.

27. By Theorem 5.12

$$\frac{BD}{DC} = \frac{AB}{AC}.$$

Also, $BC = BD + DC$ and since $BC = 45$, $45 = BD + DC$. Thus $DC = 45 - BD$. Substitute 24 for AB and 36 for AC.

$$\frac{BD}{45 - BD} = \frac{24}{36}$$
$$36DB = 24(45 - BD)$$
$$36BD = 1080 - 24BD$$
$$60BD = 1080$$
$$BD = 18$$

Then $DC = 45 - BD = 45 - 18 = 27$. Thus, $BD = 18$ in and $DC = 27$ in.

29. The figure below shows the information given.

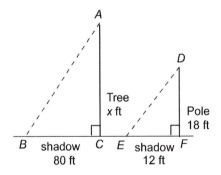

Since $\triangle ABC \sim \triangle DEF$, let x be the height of the tree, and solve:

$$\frac{x}{18} = \frac{80}{12}$$
$$12x = 18 \cdot 80$$
$$x = \frac{18 \cdot 80}{12} = 120$$

Thus, the tree is 120 ft tall.

31. *Proof:*

STATEMENTS	REASONS
1. ABC is a triangle and $\angle 1 \cong \angle 2$	1. Given
2. $\angle A \cong \angle A$	2. Reflexive law
3. $\triangle ABC \sim \triangle ADE$	3. AA
4. $\dfrac{AD}{AB} = \dfrac{DE}{BC}$	4. Def. ~ ⟨S⟩

33. *Proof:*

STATEMENTS	REASONS
1. ABC is a triangle and $\overline{DE} \parallel \overline{BC}$	1. Given
2. $\angle ADP \cong \angle ABF$ and $\angle AEP \cong \angle ACF$	2. If lines $\parallel$, corr. $\angle$'s are $\cong$
3. $\angle DAP \cong \angle BAF$ and $\angle PAE \cong \angle FAC$	3. Reflexive law
4. $\triangle ADP \sim \triangle ABF$ and $\triangle APE \sim \triangle AFC$	4. AA
5. $\dfrac{DP}{BF} = \dfrac{AP}{AF}$ and $\dfrac{AP}{AF} = \dfrac{PE}{FC}$	5. Def. ~ ⟨S⟩
6. $\dfrac{DP}{BF} = \dfrac{PE}{FC}$	6. Trans. law

35. Use Construction 5.1

37. *Proof:*

STATEMENTS	REASONS
1. Trapezoid $WXYZ$ with $\overline{XY} \parallel \overline{WZ}$	1. Given
2. $\angle YXZ \cong \angle XZW$ and $\angle XYW \cong \angle YWZ$	2. If lines $\parallel$, alt. int. $\angle$'s $\cong$
3. $\triangle XAY \sim \triangle ZAW$	3. AA

39. Answers will vary.

41. *Given:* $\triangle ABC \sim \triangle DEF$ with $\angle A \cong \angle D$.
$\overline{AP}$ and $\overline{DQ}$ are altitudes

Prove: $\dfrac{AP}{DQ} = \dfrac{AB}{DE} = \dfrac{BC}{EF} = \dfrac{AC}{DF}$

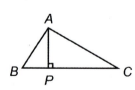
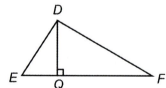

Proof: STATEMENTS   REASONS

1. $\triangle ABC \sim \triangle DEF$ with $\angle A \cong \angle D$ 1. Given

2. $\overline{AP}$ and $\overline{DQ}$ are altitudes   2. Given

3. $\overline{AP} \perp \overline{BC}$ and $\overline{DQ} \perp \overline{EF}$ 3. Def. of altitude

4. $\angle APB$ and $\angle DQE$ are right angles   4. $\perp$ lines form right $\angle$'s

5. $\angle APB \cong \angle DQE$ 5. All rt. $\angle$'s are $\cong$

6. $\angle B \cong \angle E$ 6. Corresponding $\angle$'s in $\sim$ ⧍ are $\cong$ (Def. $\sim$ ⧍)

7. $\triangle ABP \sim \triangle DEQ$ 7. AA

8. $\dfrac{AP}{DQ} = \dfrac{AB}{DE}$ 8. Corr. sides of $\sim$ ⧍ are prop. (Def. $\sim$ ⧍)

Note: Since $\dfrac{AB}{DE} = \dfrac{BC}{EF} = \dfrac{AC}{DF}$, the ratio of corresponding altitudes is equal to the ratio of <u>any</u> two corresponding sides.

43. Use Construction 5.2.

45. (a) By AA (b) $\dfrac{1}{3}$ (c) $\dfrac{1}{3}$ (d) They are 3 times longer; that is, this pantograph will enlarge a drawing to one 3 times larger than the original. (e) Place the pen at *D,* trace the original at *E* and a reduced copy will be made at *D.*

Section 5.3 Properties of Right Triangles

5.3 PRACTICE EXERCISES

1. Given **3.** $\overline{AD} \cong \overline{DB}$ thus $AD = DB$ **4.** $\triangle$ Proportionality Thm. (a line ‖ to one side and intersecting 2 sides of $\triangle$ divide the sides into proportional segments.) **5.** From Statements 3 and 4 it follows $CE = EB$ because $AD = DB$ **7.** A line ($\overline{BC}$) $\perp$ to one of 2 ‖ lines ($\overline{AC}$) is $\perp$ to the other ‖ line.

8. $\angle 1$ and $\angle 2$ are rt. $\angle$'s **10.** Reflexive law **11.** $\triangle DEB \cong \triangle DEC$ **12.** CPCTC, def. $\cong$ seg.

13. Seg.-Add. Post. **14.** Substitution (subst. *BD* for *DA* in statement 13) **15.** Dist. law

16. $2CD = BA$ (subst. *CD* for *BD* in statement 15) **17.** Mult.-Div. Post.

5.3 SECTION EXERCISES

1. $\dfrac{2}{x} = \dfrac{x}{8}$ by Cor 5.15

$x^2 = 16$

$x = 4 \text{ or } -4$

Reject -4 because a distance cannot be negative, thus $x = 4$

$\dfrac{2}{y} = \dfrac{y}{10}$ by Cor 5.15

$y^2 = 20$

$y = \pm\sqrt{20} \text{ or } \pm 2\sqrt{5}$

Reject $-\sqrt{20}$ because a distance cannot be negative, thus $y = \sqrt{20} \text{ or } 2\sqrt{5}$

3. $\dfrac{8}{x} = \dfrac{x}{8}$ by Cor 5.14

$x^2 = 64$

$x = \pm 8$

Reject -8 because a distance cannot be negative, thus $x = 8$

$\dfrac{8}{y} = \dfrac{y}{16}$ by Cor 5.15

$y^2 = 128$

$y = \pm\sqrt{128} \text{ or } \pm 8\sqrt{2}$

Reject $-\sqrt{128}$ because a distance cannot be negative thus $y = \sqrt{128} \text{ or } 8\sqrt{2}$

5. $\dfrac{x}{9} = \dfrac{9}{15}$ $y = 15 - x$

$15x = 81$ $y = 15 - \dfrac{27}{5}$

$x = \dfrac{81}{15}$ $y = \dfrac{48}{5}$

$x = \dfrac{27}{5}$

7. $\dfrac{x}{12} = \dfrac{12}{x+7}$ by Cor 5.14

$x(x+7) = 144$

$x^2 + 7x = 144$

$x^2 + 7x - 144 = 0$

$(x-9)(x+16) = 0$

$x - 9 = 0 \quad x + 16 = 0$

$x = 9 \qquad\quad x = -16$

Reject -16 because a distance cannot be negative, thus $x = 9$

9. $10 = \dfrac{1}{2}(2x)$ by Theorem 5.16

$10 = x$

11. By Theorem 5.16,

$CE = \dfrac{1}{2}AB = \dfrac{1}{2}(20) = 10 \text{ cm}$

13. By Theorem 5.14, CD is the geometric mean between AD and BD, so $\dfrac{AD}{CD} = \dfrac{CD}{BD}$.

Substitute 4 for AD and 9 for BD.

$\dfrac{4}{CD} = \dfrac{CD}{9}$

$(4)(9) = (CD)^2$ Means-extremes property

$36 = (CD)^2$

$\pm\sqrt{36} = (CD)$ Take square root of both sides

$\pm 6 = CD$

Reject -6 since a distance cannot be negative thus $CD = 6 \text{ cm}$.

15. By Theorem 5.16, $CE = \dfrac{1}{2}AB$. Thus,

$5 = \dfrac{1}{2}AB$, so $AB = 10 \text{ ft}$.

17. By Corollary 5.14, CD is the geometric mean between AD and BD, thus we have:

$$\frac{AD}{CD} = \frac{CD}{BD}$$

Substitute 3 for AD and 9 for CD.

$$\frac{3}{9} = \frac{9}{BD}$$
$$3BD = (9)(9)$$
$$3BD = 81$$
$$BD = 27$$

Thus, $BD = 27$ cm

19. By Corollary 5.15, AC is the geometric mean between AB and AD. Thus we have:

$$\frac{AB}{AC} = \frac{AC}{AD}$$

Substitute 32 for AB and 2 for AD.

$$\frac{32}{AC} = \frac{AC}{2}$$
$$(32)(2) = (AC)(AC) \quad \text{Means-extremes property}$$
$$64 = (AC)^2$$
$$\pm\sqrt{64} = AC \qquad \text{Take square root to both sides}$$
$$\pm 8 = AC$$

Reject -8 because a distance cannot be negative, thus, $AC = 8$ yd

21. By Corollary 5.15, AC is the geometric mean between AB and AD. Thus we have:

$$\frac{AB}{AC} = \frac{AC}{AD}$$

Substitute 10 for AB and 3 for AD.

$$\frac{10}{AC} = \frac{AC}{3}$$
$$(10)(3) = (AC)(AC) \quad \text{Means-extremes property}$$
$$30 = (AC)^2$$
$$\pm\sqrt{30} = AC \qquad \text{Take square root of both sides}$$

Reject $-\sqrt{30}$ because a distance cannot be negative, thus, $AC = \sqrt{30}$ ft, which can be approximated by 5.48 ft, correct to the nearest hundredth of a foot.

23. Let x be the length of AD. Then $AB = AD + BD = x + 8$. By Corollary 5.15; AC is the geometric mean between AB and AD. Thus, we must solve:

$$\frac{AB}{AC} = \frac{AC}{AD}$$
$$\frac{x+8}{7} = \frac{7}{x} \qquad \text{Substitute } x+8 \text{ for } AB,$$
$$\qquad\qquad 7 \text{ for } AC, \text{ and } x \text{ for } AD$$
$$x(x+8) = (7)(7) \quad \text{Means-extremes property}$$
$$x^2 + 8x = 49 \qquad \text{Distributive law}$$
$$x^2 + 8x - 49 = 0$$

Since the left side will not factor we use the quadratic formula with $a = 1$, $b = 8$, and $c = -49$.

$$x = \frac{-b \pm \sqrt{b^2 - 4ac}}{2a}$$
$$= \frac{(-8) \pm \sqrt{(8)^2 - 4(1)(-49)}}{2(1)}$$
$$= \frac{-8 \pm \sqrt{64 + 196}}{2}$$
$$= \frac{-8 \pm \sqrt{260}}{2}$$
$$= \frac{-8 \pm \sqrt{4 \cdot 65}}{2} = \frac{-8 \pm 2\sqrt{65}}{2}$$
$$= \frac{2(-4 \pm \sqrt{65})}{2} = -4 \pm \sqrt{65}$$

Reject $-4 - \sqrt{65}$ because a distance cannot be negative, thus, $AD = (-4 + \sqrt{65})$ cm, which can be approximated by 4.06 cm correct to the nearest hundredth of a centimeter.

25. Since $CE = 5$, and by Theorem 5.16,

$CE = \dfrac{1}{2}AB$, $5 = \dfrac{1}{2}AB$ so that $AB = 10$. By Theorem 5.15,

$$\frac{AB}{BC} = \frac{BC}{BD}.$$

Substitute 10 for AB and 8 for BD.

$$\frac{10}{BC} = \frac{BC}{8}$$
$$80 = (BC)^2 \quad \text{Means-extremes property}$$
$$\pm\sqrt{80} = BC$$
$$\pm\sqrt{16 \cdot 5} = BC$$
$$\pm 4\sqrt{5} = BC$$

Reject $-4\sqrt{5}$ because a distance cannot be negative, thus BC is $4\sqrt{5}$ yd, which can be approximated by 8.94 yd, correct to the nearest hundredth of a yard.

27. An altitude of a triangle is a line segment from a vertex perpendicular to the opposite side. A median is also a line segment from a vertex but it connects the vertex to the midpoint of the opposite side. The similarities are they are both segments from a vertex of a triangle. The difference is the altitude is always perpendicular to the opposite side and the median always bisects the opposite side.

Section 5.4 Pythagorean Theorem

5.4 PRACTICE EXERCISES

1. Use the Pythagorean Theorem $a^2 + b^2 = c^2$ with $c = 32$ and $a = 18$ and solve for b.

$$a^2 + b^2 = c^2$$
$$(18)^2 + b^2 = (32)^2$$
$$b^2 = (32)^2 - (18)^2$$
$$b^2 = 1024 - 324$$
$$b^2 = 700$$
$$b = \pm\sqrt{700} = \pm\sqrt{(100)(7)} = \pm 10\sqrt{7}$$

Thus, $b = 10\sqrt{7}$ ft (Note: only the positive value is used for length).

2. The triangle cannot be a right triangle since the longest side, 9 ft, would have to be the hypotenuse and

$$9^2 \neq 8^2 + 4^2$$

since

$$81 \neq 64 + 16 = 80$$

3. We make a sketch of the problem as shown below.

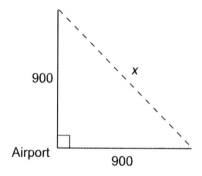

After two hours, each has traveled

$$2(450) = 900 \text{ miles.}$$

By the Pythagorean Theorem,

$$x^2 = (900)^2 + (900)^2$$
$$x^2 = 2(900)^2$$
$$x = \sqrt{2(900)^2}$$
$$x = 900\sqrt{2} \approx 1272.8 \text{ mi}$$

(Note: only the positive value is used for length)

5.4 SECTION EXERCISES

1. Substitute into the Pythagorean Theorem

$$a^2 + b^2 = c^2$$
$$3^2 + 4^2 = c^2$$
$$9 + 16 = c^2$$
$$25 = c^2$$
$$\pm\sqrt{25} = c$$
$$\pm 5 = c$$

Reject -5. Thus, $c = 5$ cm.

3. Substitute into the Pythagorean Theorem

$$a^2 + b^2 = c^2$$
$$a^2 + 25^2 = 65^2$$
$$a^2 + 625 = 4225$$
$$a^2 = 3600$$
$$a = \pm\sqrt{3600} = \pm 60$$

Reject -60. Thus, $a = 60$ ft.

5. Substitute into the Pythagorean Theorem

$$a^2 + b^2 = c^2$$
$$6^2 + b^2 = 11^2$$
$$36 + b^2 = 121$$
$$b^2 = 85$$
$$b = \pm\sqrt{85}$$

Reject $-\sqrt{85}$. Thus, $b = \sqrt{85}$ yd.

7. Substitute into the Pythagorean Theorem

$$a^2 + b^2 = c^2$$
$$8^2 + b^2 = (2\sqrt{97})^2$$
$$64 + b^2 = 4(97) = 388$$
$$b^2 = 324$$
$$b = \pm\sqrt{324} = \pm 18$$

Reject -18. Thus, $b = 18$ cm.

9. Yes, by the converse of the Pythagorean Theorem, $15^2 + 20^2 = 225 + 400 = 625 = 25^2$.

11. Yes, by the converse of the Pythagorean Theorem, $3^2 + 7^2 = 9 + 49 = 58 = (\sqrt{58})^2$.

13. No, since $(\sqrt{7})^2 + (\sqrt{2})^2 = 7 + 2 = 9 \neq 81 = 9^2$.

15. By Theorem 5.19,

$$c = \sqrt{2}a = (\sqrt{2})(10)$$
$$c = 10\sqrt{2} \text{ ft}$$

17. Since $b = a$ and $a = 3\sqrt{2}$ yd, $b = 3\sqrt{2}$ yd.

19. By Theorem 5.19,

$$c = \sqrt{2}a = \sqrt{2}(3\sqrt{2})$$
$$= 3(\sqrt{2})(\sqrt{2}) = 3(2) = 6 \text{ cm.}$$

21. By Theorem 5.19,

$$c = \sqrt{2}b = \sqrt{2}(3\sqrt{3})$$
$$= 3\sqrt{2}\sqrt{3} = 3\sqrt{6} \text{ ft.}$$

23. By Theorem 5.19, $c = \sqrt{2}a$. Substitute 6 for c and solve for a: $6 = \sqrt{2}a$ so that

$$a = \frac{6}{\sqrt{2}} = \frac{6}{\sqrt{2}} \cdot \frac{\sqrt{2}}{\sqrt{2}} = \frac{6\sqrt{2}}{2} = 3\sqrt{2} \text{ yd.}$$

25. By Theorem 5.19, $c = \sqrt{2}b$. Substitute $\frac{\sqrt{2}}{2}$ for c and solve for b. $\frac{\sqrt{2}}{2} = \sqrt{2}b$ so that $b = \frac{\sqrt{2}}{2} \cdot \frac{1}{\sqrt{2}} = \frac{1}{2}$ cm.

27. By Theorem 5.20, $b = \frac{1}{2}c$ so $10 = \frac{1}{2}c$ making $c = 20$ ft.

29. By Theorem 5.20, $b = \frac{1}{2}c$ so $b = \frac{1}{2}(16) = 8$ yd.

31. By Theorem 5.20,

$$a = \sqrt{3}b = \sqrt{3}(7) = 7\sqrt{3} \text{ cm}$$

33. By Theorem 5.20,

$$a = \sqrt{3}b \text{ so that } 2\sqrt{3} = \sqrt{3}b.$$

Dividing both sides by $\sqrt{3}$ gives $b = 2$ ft.

35. By Theorem 5.20,

$$a = \frac{\sqrt{3}}{2}c = \frac{\sqrt{3}}{2}(\sqrt{3}) = \frac{\sqrt{3}}{2}\frac{\sqrt{3}}{2} \cdot \frac{3}{2} = \frac{3}{2} \text{ yd}$$

37. By Theorem 5.20, $a = \dfrac{\sqrt{3}}{2}c.$

Thus, $\sqrt{3} = \dfrac{\sqrt{3}}{2}c.$

Multiply both sides by

$$\frac{2}{\sqrt{3}} : \frac{2}{\sqrt{3}}(\sqrt{3}) = \frac{2}{\sqrt{3}} \cdot \frac{\sqrt{3}}{2}c.$$

Thus, $c = 2$ cm.

39. The diagonal, d, of a square with sides 4 inches long is the hypotenuse of a right triangle with legs 4 inches. By the Pythagorean Theorem,

$$d^2 = 4^2 + 4^2 = 16 + 16 = 32$$
$$d = \pm\sqrt{32} = \pm\sqrt{16 \cdot 2} = \pm 4\sqrt{2}$$

Reject the negative value $-4\sqrt{2}$. Thus, the diagonal is $4\sqrt{2}$ inches.

41. Consider the figure below.

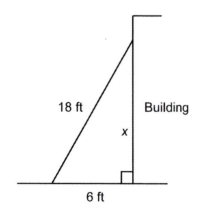

Use the Pythagorean Theorem to find x.

$$x^2 + 6^2 = 18^2$$
$$x^2 + 36 = 324$$
$$x^2 = 288$$
$$x = \pm\sqrt{288} = \pm\sqrt{144 \cdot 2} = \pm 12\sqrt{2}$$

Reject the negative value $-12\sqrt{2}$. Thus, the ladder will reach $12\sqrt{2}$ ft, approximately 17.0 ft, up the side of the building.

43.   Consider the figure below.

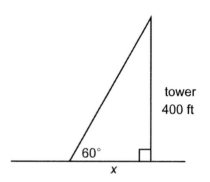

Since the triangle formed is a $30°\text{-}60°\text{-}90°$ triangle, the side opposite the $30°$ angle, x, is given in terms of the side opposite the $60°$-angle by

$$400 = \sqrt{3}x.$$

Divide both sides by $\sqrt{3}$.

$$x = \frac{400}{\sqrt{3}} = \frac{400}{\sqrt{3}} \cdot \frac{\sqrt{3}}{\sqrt{3}} = \frac{400\sqrt{3}}{3} \approx 230.94011$$

Thus, to the nearest tenth of a foot, the wire is 230.9 ft from the base of the tower.
(Note: only the positive value is used for length)

45. Consider the equilateral triangle with sides 10 ft shown below. Each angle measures $60°$.

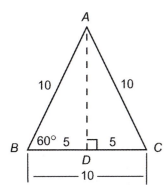

Draw an altitude from A to $\overline{BC}$. Then $\triangle ABD$ is a $30°\text{-}60°\text{-}90°$ triangle. By Theorem 5.20, $AD = \sqrt{3}BD = \sqrt{3}(5) = 5\sqrt{3}$.

Thus, the altitude of the triangle is $5\sqrt{3}$ ft. (Note: only the positive value is used for length)

47. Consider the equilateral triangle with sides 10 ft shown below. Each angle measures $60°$.

Draw a perpendicular from A to $\overline{BC}$. Then $\triangle ABD$ is a $30°\text{-}60°\text{-}90°$ triangle. By Theorem 5.20, $AD = BD\sqrt{3} = 5\sqrt{3}$. Then the height of $\triangle ABC$ is $AD = 5\sqrt{3}$ and its base is $BC = 10$. Thus,

$$A = \frac{1}{2}bh = \frac{1}{2}(10)(5\sqrt{3}) = 25\sqrt{3} \text{ ft}^2.$$

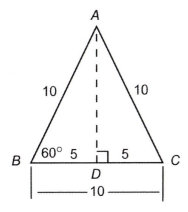

49. Find the length of diagonal d in the cube with sides x shown below.

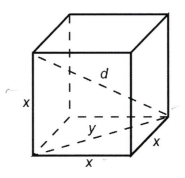

First draw diagonal of the base y. Then y can be found by the Pythagorean Theorem.

$$x^2 + x^2 = y^2$$
$$2x^2 = y^2$$
$$\sqrt{2}x = y$$

Now find d using the Pythagorean Theorem again.

$$x^2 + y^2 = d^2$$
$$x^2 + (\sqrt{2}x)^2 = d^2$$
$$x^2 + 2x^2 = d^2$$
$$3x^2 = d^2$$
$$\sqrt{3}x = d$$

Note we only use the positive root in each case. Thus, diagonal d has length $\sqrt{3}x$ where x is the length of each side of the cube.

51. $\sqrt{3}$ times the length of the original segment. The process can be continued to find a segment $\sqrt{n}$ times the length of the original segment for $n = 2, 3, 4, 5, \ldots$.

53. The quadrilateral is clearly a rhombus, because all sides have same length. Show it contains a right angle by showing the angle is the supplement of the angle formed by adding the two acute angles of the right triangle.

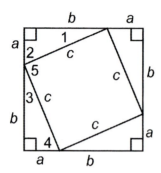

The outside figure is given to be a square.

Thus 4 rt. are formed and from the given lengths of the sides,, they are $\cong$ by SSS. Thus corr. $\angle$'s $\cong$, $\angle 1 \cong \angle 3$, $\angle 2 \cong \angle 4$.

$m\angle 1 + m\angle 2 = 90°$ because the acute $\angle$'s of a rt. $\triangle$ must sum to 90° because the sum of $\angle$'s $\triangle = 180°$.

Using substitution $m\angle 3 + m\angle 2 = 90°$.

$m\angle 2 + m\angle 3 + m\angle 5 = 180°$ because their exterior sides lie in a straight line. Thus $m\angle 5 = 90°$. Use a similar argument for the other $\angle$'s. Thus, the "inside" figure is a square.

55. (a) The 2 rt. $\triangle$ are $\cong$ by SSS $(\triangle AED \cong \triangle BCA)$ therefore

$\angle CAB \cong \angle EDA$ and $\angle CBA \cong \angle EAD$ by CPCTC

$m\angle CAB + m\angle CBA = 90°$ because the acute $\angle$'s of rt. $\triangle$ sum to 90° since the sum of measures of all $\angle$'s$\triangle = 180°$.

$m\angle CAB + m\angle EAD = 90°$ by substituting

$m\angle CAB + m\angle EAD + m\angle DAB = 180°$ since their exterior sides lie in a straight line.

$90° + m\angle DAB = 180°$ by substitution, thus $m\angle DAB = 90°$.

(b) $\frac{1}{2}ab$, $\frac{1}{2}ab$, $\frac{1}{2}c^2$, since the 3 $\triangle$ are rt. $\triangle$

area = $\frac{1}{2}$ the product of the legs because the legs are a base and height.

(c) $\overline{ED} \parallel \overline{CB}$ because both lines are perpendicular to the same line. Thus $BCED$ is a quadrilateral with one pair of parallel sides.

(d) Area trapezoid = $\frac{1}{2}h(b_1 + b_2)$ where $h =$ height, b_1 and b_2 are the bases (parallel sides).

$$A = \frac{1}{2}(EC)(ED + CB) = \frac{1}{2}(a+b)(a+b)$$

(e) Area trapezoid = sum area of 3 triangles

$$\frac{1}{2}(a+b)(a+b) = \frac{1}{2}ab + \frac{1}{2}ab + \frac{1}{2}c^2$$

$$\frac{1}{2}(a^2 + 2ab + b^2) = ab + \frac{1}{2}c^2$$

$$\frac{1}{2}a^2 + ab + \frac{1}{2}b^2 = ab + \frac{1}{2}c^2 \quad \text{Subt. } ab \text{ from both sides of equation.}$$

$$\frac{1}{2}a^2 + \frac{1}{2}b^2 = \frac{1}{2}c^2 \quad \text{Mult. by 2}$$

$$a^2 + b^2 = c^2$$

Section 5.5 Inequalities Involving Triangles (optional)

5.5 PRACTICE EXERCISES

1.

$$5(2x-1) > 12x+7$$
$$10x-5 > 12x+7 \qquad \text{Distributive law}$$
$$10x-5+5 > 12x+7+5 \qquad \text{Add 5 to both sides}$$
$$10x > 12x+12 \qquad \text{Simplify}$$
$$10x-12x > 12x-12x+12 \quad \text{Subtract } 12x \text{ from both sides}$$
$$-2x > 12 \qquad \text{Simplify}$$
$$\frac{-2x}{-2} < \frac{12}{-2} \qquad \text{Divide both sides by } -2 \text{ and reverse inequality sign}$$
$$x < -6$$

2. In $\triangle PQR$, if $m\angle P = 75°$ and $m\angle Q = 65°$, then $m\angle R = 180° - m\angle P - m\angle Q = 180° - 75° - 65° = 40°$. The shortest side is opposite the smallest angle, which is $\angle R$, so the shortest side is $\overline{PQ}$. The longest side is opposite the largest angle, which is $\angle P$, so the longest side is $\overline{QR}$.

3. By the SAS Inequality Theorem, since $AB = DE$ and $BC = EF$, with $m\angle B = 36° < 42° = m\angle E$, side $\overline{AC}$ is shorter than side $\overline{DF}$. Thus, $DF > AC$.

5.5 SECTION EXERCISES

1. Transitive law

3. Division property of inequalities

5. Subtraction property of inequalities

7. The whole is greater than its parts

9. Transitive law

11.
$$x+3 < 7$$
$$x+3-3 < 7-3 \quad \text{Subtraction property of inequalities}$$
$$x < 4 \qquad \text{Simplify}$$

13.
$$1-3x < 8$$
$$1-1-3x < 8-1 \quad \text{Subtraction property of inequalities}$$
$$-3x < 7 \qquad \text{Simplify}$$
$$\frac{-3x}{-3} > \frac{7}{-3} \qquad \text{Division property of inequalities (Note reversal of sign)}$$
$$x > -\frac{7}{3} \qquad \text{Simplify}$$

15.
$$\frac{2x-1}{3} < 5$$
$$3\left(\frac{2x-1}{3}\right) < (3)(5) \quad \text{Multiplication property of inequalities}$$
$$2x-1 < 15 \qquad \text{Simplify}$$
$$2x-1+1 < 15+1 \qquad \text{Addition property of inequalities}$$
$$2x < 16 \qquad \text{Simplify}$$
$$\frac{2x}{2} < \frac{16}{2} \qquad \text{Division property of inequalities}$$
$$x < 8 \qquad \text{Simplify}$$

17.

$$3(2x+8) < 4(x-3)$$

$6x+24 < 4x-12$	Distributive law
$6x+24-24 < 4x-12-24$	Subtraction property of inequalities
$6x < 4x-36$	Simplify
$6x-4x < 4x-4x-36$	Subtraction property of inequalities
$2x < -36$	Simplify
$\dfrac{2x}{2} < \dfrac{-36}{2}$	Division property of inequalities
$x < -18$	Simplify

19. False; an exterior angle of a triangle is greater than (not equal to) each remote interior angle.

21. True; an exterior angle of a triangle is greater than each remote interior angle.

23. True; if two sides of a triangle are unequal, the angles opposite these sides are unequal in the same order.

25. False; $BC > 16$ yd since sides opposite equal angles are unequal in the same order.

27. True; sides of a triangle are unequal in the same order as the unequal angles opposite them.

29. True; by the Triangle Inequality, the sum of the lengths of any two sides of a triangle $(AB + AC = 10 + 7 = 17)$ must be greater than the third side (BC).

31. $AC > DE$ by the SAS Inequality Theorem (Theorem 5.25).

33. $m\angle C > m\angle D$ since $AB = 7 > EF = 6$, $\angle C$ must be greater than $\angle D$ by the SSS Inequality Theorem (Theorem 5.26).

35. The Triangle Inequality prevents this from happening since $3 + 4 = 7 < 8$ not greater than 8.

37. *Proof:*

STATEMENTS	REASONS
1. $BQ = ED$ thus $\overline{BQ} \cong \overline{ED}$	1. By Construction; Def. $\cong$ seg
2. $BC = EF$ thus $\overline{BC} \cong \overline{EF}$	2. Given; Def. $\cong$ seg
3. $m\angle QBC = m\angle DEF$ thus $\angle QBC \cong \angle DEF$	3. Given; Def. $\cong$ $\angle$'s.
4. $\triangle QBC \cong \triangle DEF$	4. SAS
5. $\overline{QC} \cong \overline{DF}$ thus $QC = DF$	5. CPCTC; Def. $\cong$ seg
6. $AC = QC + AQ$	6. Seg.-Add. Post.
7. $AC > QC$	7. Whole > any part
8. $AC > DF$	8. Substitution law

39. *Proof:*

STATEMENTS	REASONS
1. $AB = DE$ and $BC = EF$ thus $\overline{AB} \cong \overline{DE}$; $\overline{BC} \cong \overline{EF}$	1. Given; Def. $\cong$ seg
2. $BQ = ED$ thus $\overline{BQ} \cong \overline{ED}$	2. By construction; def. $\cong$ seg
3. $\overline{AB} \cong \overline{BQ}$	3. Trans. laws and statements 1 and 2
4. $\overline{BR} \cong \overline{BR}$	4. Reflexive law
5. $\overline{BR}$ bisects $\angle ABQ$	5. By construction
6. $\angle ABR \cong \angle RBQ$	6. Def. of $\angle$ bisector
7. $\triangle ABR \cong \triangle RBQ$	7. SAS
8. $\overline{AR} \cong \overline{RQ}$ thus $AR = RQ$	8. CPCTC; Def. $\cong$ seg
9. $AC = AR + RC$	9. Seg.-Add. Post.
10. $AC = RQ + RC$	10. Substitution law
11. $RQ + RC > QC$	11. Triangle Inequality Theorem
12. $AC > QC$	12. Substitution law
13. $\angle QBC \cong \angle DEF$	13. By construction
14. $\triangle QBC \cong \triangle DEF$	14. SAS
15. $\overline{QC} \cong \overline{DF}$ thus $QC = DF$	15. CPCTC; Def. $\cong$ seg
16. $AC > DF$	16. Substitution law

41. *Proof:*

STATEMENTS	REASONS
1. $m\angle A > m\angle B$	1. Given
2. $BC > AC$	2. Sides opp. $\neq \angle$'s are $\neq$ in same order
3. $m\angle D > m\angle E$	3. Given
4. $CE > CD$	4. Same as statement 2
5. $BC + CE > AC + CD$	5. Add. prop. of inequalities
6. $BE = BC + CE$ and $AD = AC + CD$	6. Seg.- Add. Post.
7. $BE > AD$	7. Substitution law

43. *Given:* $\triangle ABC$

Prove: $AC - AB < BC$

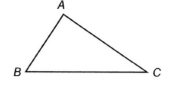

Proof: STATEMENTS   REASONS

    1. ABC is a triangle   1. Given

    2. $AC < AB + BC$ 2. Triangle Inequality

    3. $AB = AB$ 3. Reflexive law

    4. $AC - AB < BC$ 4. Subtraction prop. of inequalities
                                                     using statement 2 and 3

45. *Given:* $\triangle ABC$ with P inside the triangle

Prove: $PA + PB + PC > \dfrac{1}{2}(AB + BC + AC)$

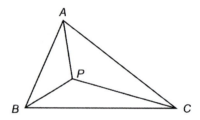

Proof: STATEMENTS   REASONS

    1. ABC is a triangle with P inside it   1. Given

    2. $PA + PB > AB, PA + PC > AC$, and   2. Triangle Inequality applied to
       $PB + PC > BC$ $\triangle APB, \triangle APC,$ and $\triangle BPC$

    3. $PA + PB + PA + PC + PB + PC > AB + AC + BC$ 3. Add. prop. of inequalities

    4. $2PA + 2PB + 2PC > AB + AC + BC$ 4. Distributive law

    5. $2(PA + PB + PC) > AB + AC + BC$ 5. Distributive law

    6. $PA + PB + PC > \dfrac{1}{2}(AB + AC + BC)$ 6. Div. prop. of inequalities

47. *Proof:*

	STATEMENTS		REASONS
	1. $PA = AD$		1. D is $\perp$ reflection of P in mirror
	2. $AB = AB$, $AC = AC$		2. Reflexive law
	3. $\angle DAC$ and $\angle PAC$ are right angles		3. $\perp$ lines form rt. $\angle$'s
	4. $\triangle DAC$, $\triangle PAC$, $\triangle DAB$, and $\triangle PAB$ are right triangles		4. Def. of rt $\triangle$
	5. $\triangle DAB \cong \triangle PAB$ and $\triangle DAC \cong \triangle PAC$		5. LL
	6. $\overline{DB} \cong \overline{PB}$ and $\overline{DC} \cong \overline{PC}$ thus $DB = PB$ and $DC = PC$		6. CPCTC; Def. $\cong$ seg.
	7. $DQ = DB + BQ$		7. Seg.-Add. Post.
	8. $DQ < DC + CQ$		8. Triangle Inequality
	9. $DB + BQ < DC + CQ$		9. Substitution law
	10. $PB + BQ < PC + CQ$		10. Substitution law using statements 6 and 9

49. Consider the figure below showing the given information.

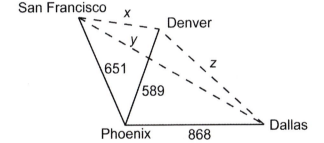

San Francisco x Denver

651

y

589

z

Phoenix 868 Dallas

(a) If x is the air distance from Denver to San Francisco, then by using the Triangle Inequality three times we have:

$x < 651 + 589$; $651 < x + 589$; $589 < x + 651$
$x < 1240$; $62 < x$; $-62 < x$

Since $x > 62$ certainly includes $x > -62$, we have that
$$62 < x < 1240.$$

The minimum distance is 62 miles and the maximum distance is 1240 miles.

(b) If y is the air distance from San Francisco to Dallas, then by using the Triangle Inequality three times we have:

$y < 651 + 868$; $868 < 651 + y$; $651 < 868 + y$
$y < 1519$; $217 < y$; $-217 < y$

Since $y > 217$ certainly includes $y > -217$, we have that
$$217 < y < 1519.$$

The minimum distance is 217 miles and the maximum distance is 1519 miles.

(c) If z is the air distance from Denver to Dallas, then by using the Triangle Inequality three times we have:

$z < 589 + 868$; $868 < 589 + z$; $589 < 868 + z$
$z < 1457$; $279 < z$; $-279 < z$

Since $z > 279$ includes $z > -279$, we have that
$$279 < z < 1457.$$

The minimum distance is 279 miles and the maximum distance is 1457 miles.

Chapter 5 Review Exercises

1. $\dfrac{32}{40} = \dfrac{4 \cdot \cancel{8}}{5 \cdot \cancel{8}} = \dfrac{4}{5}$

2. $\dfrac{300 \text{ mi}}{5 \text{ hr}} = \dfrac{\cancel{5} \cdot 60}{\cancel{5}} \dfrac{\text{mi}}{\text{hr}} = 60 \dfrac{\text{mi}}{\text{hr}} = 60 \text{ mph}$

3. $\dfrac{a}{12} = \dfrac{1}{4}$

$4a = (12)(1)$

$4a = 12$

$a = \dfrac{12}{4} = 3$

4. $\dfrac{4}{20} = \dfrac{8}{x}$

$4x = (8)(20)$

$x = \dfrac{(8)(20)}{4} = (2)(20) = 40$

5. $\dfrac{y+5}{y} = \dfrac{21}{6}$

$6(y+5) = 21y$

$6y + 30 = 21y$

$30 = 15y$

$2 = y$

6. $\dfrac{a+3}{36} = \dfrac{a-3}{9}$

$9(a+3) = 36(a-3)$

$9a + 27 = 36a - 108$

$27 = 27a - 108$

$135 = 27a$

$5 = a$

7. 6 is to 5 as 24 is to x translates to:

$\dfrac{6}{5} = \dfrac{24}{x}$

$6x = (5)(24)$

$x = \dfrac{(5)(24)}{6} = (5)(4) = 20$

8. Let x be the mean proportional or geometric mean between 16 and 25. We must solve:

$$\dfrac{16}{x} = \dfrac{x}{25}$$
$$(16)(25) = (x)(x)$$
$$400 = x^2$$

Since $(20)^2 = 400$ and $(-20)^2 = 400$, x is 20 or x is -20.

9. The problem translates to the following proportion

$$\dfrac{\frac{1}{4}}{20} = \dfrac{2\frac{1}{4}}{x},$$

where x is the number of miles represented by $2\dfrac{1}{4}$ inches. Then

$$\dfrac{1}{4}x = (20)\left(2\dfrac{1}{4}\right)$$
$$\dfrac{1}{4}x = (20)\left(\dfrac{9}{4}\right)$$
$$\dfrac{1}{4}x = (5)(9)$$
$$\dfrac{1}{4}x = 45$$
$$x = (4)(45)$$
$$x = 180$$

Thus $2\dfrac{1}{4}$ inches on the map represents an actual distance of 180 miles.

10. False; $ad = bc$ but $ac \neq bd$

11. True

12. True

13. $\dfrac{y}{7} = \dfrac{4}{8}$

$8y = (4)(7)$

$y = \dfrac{(4)(7)}{8} = \dfrac{7}{2} = 3.5$ cm

14. $\dfrac{x}{4} = \dfrac{4}{8}$

$8x = (4)(4)$

$x = \dfrac{(4)(4)}{8} = 2$ cm

15. $\dfrac{z}{4} = \dfrac{8}{4}$

$4z = (4)(8)$

$z = \dfrac{(4)(8)}{4} = 8$ cm

16. Since the sum of the angles of a quadrilateral is

$S = (4-2)(180°) = (2)(180°) = 360°,$

$m\angle A + m\angle B + m\angle C + m\angle D = 360°$
$75° + 76° + m\angle C + 114° = 360°$
$m\angle C + 265° = 360°$
$m\angle C = 360° - 265° = 95°$

Since $m\angle C' = m\angle C, m\angle C' = 95°$.

17. Yes; by AA

18. No; all $\angle$'s in equilateral $\triangle$ measure 60°

19. $\dfrac{AD}{DB} = \dfrac{AE}{EC}$

Substitute 20 for *AD*, 15 for *DB*, and 28 for *AE*.

$\dfrac{20}{15} = \dfrac{28}{EC}$
$20EC = (15)(28)$
$EC = \dfrac{(15)(28)}{20} = 21$

Thus, $EC = 21$ ft.

20. Since $AB = AD + DB$, $48 = AD + 12$, so $AD = 48 - 12 = 36$. Since $\triangle ADE \sim \triangle ABC$,

$\dfrac{AD}{AB} = \dfrac{AE}{AC}$
$\dfrac{36}{48} = \dfrac{AE}{50}$
$(36)(50) = 48AE$
$\dfrac{(36)(50)}{48} = AE$
$37.5 = AE$

Since $AC = AE + EC, 50 = 37.5 + EC$, so $EC = 50 - 37.5 = 12.5$. Thus, $AE = 37.5$ ft and $EC = 12.5$ ft.

21. By the Triangle angle-bisector theorem,

$\dfrac{BF}{FC} = \dfrac{AB}{AC}.$
$\dfrac{14}{FC} = \dfrac{35}{50}$
$(14)(50) = 35FC$
$\dfrac{(14)(50)}{35} = FC$
$20 = FC$

Thus, $FC = 20$ cm.

22. Since $BC = BF + FC = 77$, $BF = 77 - FC$. By the Triangle angle-bisector theorem,

$\dfrac{BF}{FC} = \dfrac{AB}{AC}$
$\dfrac{77 - FC}{FC} = \dfrac{30}{80}$
$80(77 - FC) = 30FC$
$6160 - 80FC = 30FC$
$6160 = 110FC$
$56 = FC$

Then $BF = 77 - EC = 77 - 56 = 21$. Thus, $BF = 21$ ft and $FC = 56$ ft.

23. *Proof:*

STATEMENTS	REASONS
1. $\triangle ABC$ is isosceles with base $\overline{BC}$	1. Given
2. $\angle C$ is supplementary to $\angle 1$	2. Given
3. $m\angle C + m\angle 1 = 180°$	3. Def. supp. angles
4. $\angle C \cong \angle B$; thus $m\angle C = m\angle B$	4. Base angles of isos. $\triangle \cong$; def. $\cong \angle$'s
5. $m\angle B + m\angle 1 = 180°$	5. Substitution
6. $\angle B$ supp. to $\angle 1$	6. Def. supp. angles
7. $\overline{DE} \parallel \overline{BC}$	7. If interior $\angle$'s on same side of transversal supp., then lines $\parallel$.
8. $\dfrac{AD}{DB} = \dfrac{AE}{EC}$	8. Line $\parallel$ to third side of $\triangle$ divides other sides into prop. segments (Thm 5.11)

24. *Proof:*

STATEMENTS	REASONS
1. ABC is a triangle with $\overline{BD}$ bisecting $\angle B$	1. Given
2. $\overline{ED} \parallel \overline{BC}$	2. Given
3. $\dfrac{AE}{EB} = \dfrac{AD}{DC}$	3. Line $\parallel$ to 3rd side of $\triangle$ divides other sides in prop. seg. (Theorem 5.11)
4. $\dfrac{AB}{BC} = \dfrac{AD}{DC}$	4. Bisector of $\angle$ divides side into seg. prop. to other two sides (Theorem 5.12)
5. $\dfrac{AE}{EB} = \dfrac{AB}{BC}$	5. Trans. law

25. Use Construction 5.1.

26. By Theorem 5.16,

$$CE = \frac{1}{2} AB = \frac{1}{2}(44) = 22 \text{ cm.}$$

27. By Corollary 5.14,

$$\frac{AD}{CD} = \frac{CD}{DB}.$$

Substitute 25 for AD and 16 for DB.

$$\frac{25}{CD} = \frac{CD}{16}$$
$$(25)(16) = (CD)^2$$
$$400 = (CD)^2$$
$$\pm\sqrt{400} = CD$$
$$\pm 20 = CD$$

Reject -20 because a length cannot be negative. Thus $CD = 20$ ft.

28. By Corollary 5.15,

$$\frac{AB}{AC} = \frac{AC}{AD}.$$

Substitute 28 for AB and 16 for AD.

$$\frac{28}{AC} = \frac{AC}{16}$$
$$(28)(16) = (AC)^2$$
$$448 = (AC)^2$$
$$\pm\sqrt{448} = AC$$
$$\pm\sqrt{64 \cdot 7} = AC$$
$$\pm 8\sqrt{7} = AC$$

Reject the negative value, $-8\sqrt{7}$. Thus, $AC = 8\sqrt{7}$ cm.

29. Let $x = AD$. By Corollary 5.15,

$$\frac{AB}{AC} = \frac{AC}{AD}$$

Substitute $x + 5$ for AB ($AB = AD + BD$), 6 for AC, and x for AD.

$$\frac{x+5}{6} = \frac{6}{x}$$
$$x(x+5) = (6)(6)$$
$$x^2 + 5x = 36$$
$$x^2 + 5x - 36 = 0$$
$$(x-4)(x+9) = 0$$
$$x - 4 = 0 \text{ or } x + 9 = 0$$
$$x = 4 \text{ or } \qquad x = -9$$

Reject the negative value, -9. Thus, $AD = x = 4$ yd.

30. First find the height of $\triangle ABC$, CD, by using Corollary 5.14,

$$\frac{AD}{CD} = \frac{CD}{DB}$$

Substitute 36 for AD and 25 for DB.

$$\frac{36}{CD} = \frac{CD}{25}$$
$$(36)(25) = (CD)^2$$
$$900 = (CD)^2$$
$$\pm\sqrt{900} = CD$$
$$\pm 30 = CD$$

Reject -30, so $CD = 30$ ft. The area of $\triangle ABC$ is given by

$$A = \frac{1}{2}(AB)(CD)$$
$$= \frac{1}{2}(61)(30) \quad AB = AD + DB = 36 + 25 = 61$$
$$= (61)(15)$$
$$= 915 \text{ ft}^2$$

31. *Given:* $ABCD$ is a rhombus with diagonals
                 $\overline{AC}$ and $\overline{BD}$, and $\overline{PQ} \perp \overline{BC}$

Prove: $(PE)^2 = (BP)(PC)$

Proof: STATEMENTS   REASONS

1. $ABCD$ is a rhombus with diagonals $\overline{AC}$ and $\overline{BD}$.   1. Given

2. $\overline{AC} \perp \overline{BD}$ 2. Diag. of rhombus are $\perp$

3. $\angle BEC$ is a right angle   3. $\perp$ lines form rt. $\angle$'s

4. $\triangle BEC$ is a right triangle   4. Def. of rt $\triangle$

5. $\overline{PQ} \perp \overline{BC}$ 5. Given

6. $\overline{EP}$ is an altitude of $\triangle BEC$ from right angle $\angle BEC$ 6. Def. of altitude

7. $\dfrac{BP}{PE} = \dfrac{PE}{PC}$ 7. Corollary 5.14

8. $(PE)^2 = (BP)(PC)$ 8. Means-extremes prop. (Thm 5.1)

32. *Given:* $ABCD$ is a rhombus with diagonals
                 $\overline{AC}$ and $\overline{BD}$, and $\angle EQA$ is a right
                 angle.

Prove: $(AE)^2 = (AQ)(AD)$

Proof: STATEMENTS   REASONS

1. $ABCD$ is a rhombus with diagonals $\overline{AC}$ and $\overline{BD}$ 1. Given

2. $\angle EQA$ is a right angle   2. Given

3. $\overline{EQ} \perp \overline{AD}$ 3. Lines meeting in rt. $\angle$ are $\perp$

4. $\overline{BD} \perp \overline{AC}$ 4. Diag. of rhombus are $\perp$

5. $\angle AED$ is a right angle   5. $\perp$ lines form rt. $\angle$'s

6. $\triangle AED$ is a right triangle   6. Def. of rt. $\triangle$

7. $\overline{EQ}$ is an altitude of $\triangle AED$ from right angle $\angle AED$ 7. Def. of alt.

8. $\dfrac{AD}{AE} = \dfrac{AE}{AQ}$ 8. Corollary 5.15

9. $(AE)^2 = (AD)(AQ)$ 9. Means-extremes prop.
                                                                  (Thm 5.1)

33. Use the Pythagorean Theorem to find the other leg x.

$$x^2 + 80^2 = 100^2$$
$$x^2 + 6400 = 10,000$$
$$x^2 = 3600$$
$$x = \pm\sqrt{3600} = \pm 60$$

Reject $x = -60$. Thus, $x = 60$. Since the legs are perpendicular, one leg can be thought of as the height and the other as the base of the triangle. Thus,

$$A = \frac{1}{2}bh = \frac{1}{2}(60)(80) = (30)(80) = 2400 \text{ yd}^2.$$

34. Use the Pythagorean Theorem.

$$a^2 + b^2 = c^2$$
$$11^2 + 8^2 = c^2$$
$$121 + 64 = c^2$$
$$185 = c^2$$
$$\pm\sqrt{185} = c$$

Reject the negative value $-\sqrt{185}$. Thus, $c = \sqrt{185}$ cm.

35. Use the Pythagorean Theorem.

$$a^2 + b^2 = c^2$$
$$14^2 + b^2 = (2\sqrt{170})^2$$
$$196 + b^2 = 4(170) = 680$$
$$b^2 = 484$$
$$b = \pm\sqrt{484} = \pm 22$$

Reject the negative value, -22. Thus, $b = 22$ ft.

36. No; since

$$6^2 + (\sqrt{11})^2 = 36 + 11 = 47 \neq 49 = 7^2$$

37. By Theorem 5.20, $a = \frac{1}{2}c$. Thus, $30 = \frac{1}{2}c$ so that $c = 60$ cm.

38. By Theorem 5.20, $b = \sqrt{3}a$. Thus, $b = \sqrt{3}(20) = 20\sqrt{3}$ yd.

39. By Theorem 5.20, $a = \frac{1}{2}c$. Thus,

$$a = \frac{1}{2}(8) = 4 \text{ ft.}$$

40. By Theorem 5.20, $b = \frac{\sqrt{3}}{2}c$. Thus,

$$b = \frac{\sqrt{3}}{2}(12) = \sqrt{3}(6) = 6\sqrt{3} \text{ cm.}$$

41. By Theorem 5.20, $b = \sqrt{3}a$. Thus, $3\sqrt{3} = \sqrt{3}a$. Divide both sides by $\sqrt{3}$ to obtain $a = 3$ yd.

42. By Theorem 5.20, $b = \frac{\sqrt{3}}{2}c$. Thus, $7 = \frac{\sqrt{3}}{2}c$. Multiply both sides by $\frac{2}{\sqrt{3}}$. $7\left(\frac{2}{\sqrt{3}}\right) = c$

making $c = \frac{14}{\sqrt{3}} = \frac{14\sqrt{3}}{\sqrt{3}\sqrt{3}} = \frac{14\sqrt{3}}{3}$ ft.

43. By Theorem 5.19, $c = \sqrt{2}\,d$. Thus, $\frac{\sqrt{2}}{5} = \sqrt{2}d$. Divide both sides by $\sqrt{2}$ to obtain $d = \frac{1}{5}$ yd.

44. By Theorem 5.19, $c = \sqrt{2}d$. Thus, $c = \sqrt{2}(5\sqrt{2}) = 5\sqrt{2}\sqrt{2} = 5(2) = 10$ cm.

45. By Theorem 5.20, $a = \dfrac{1}{2}c$ so $3 = \dfrac{1}{2}c$ making

$c = 6$ ft. By Theorem 5.19, $c = \sqrt{2}d$ so
$6 = \sqrt{2}d$. Divide both sides by $\sqrt{2}$. Then

$$d = \frac{6}{\sqrt{2}} = \frac{6}{\sqrt{2}} \cdot \frac{\sqrt{2}}{\sqrt{2}} = \frac{6\sqrt{2}}{2} = 3\sqrt{2} \text{ ft.}$$

46. The diagonal, d, of the rectangle is the
hypotenuse of a right triangle with legs 18 ft
and 7 ft. Use the Pythagorean Theorem.

$$d^2 = 18^2 + 7^2$$
$$d^2 = 324 + 49$$
$$d^2 = 373$$
$$d = \pm\sqrt{373} \approx \pm 19.313208$$

Reject the negative value so that $d = 19.3$ ft,
correct to the nearest tenth of a foot.

47. Consider the figure below.

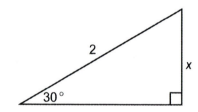

We must find x. Since $x = \dfrac{1}{2}$(hypotenuse),

$x = \dfrac{1}{2}(2) = 1$. Thus, the truck will change

altitude in the amount of 1 mile when driving
2 miles on the road.

48. Consider the equilateral triangle with sides
x shown below.

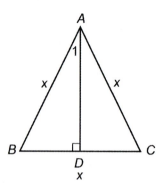

Construct an altitude. Since the altitude
bisects the vertex angle,
$60°$, $m\angle 1 = 30°$, $m\angle B = 60°$, and
$m\angle ADB = 90°$. By Theorem 5.20
$AD = \dfrac{\sqrt{3}}{2} AB = \dfrac{\sqrt{3}}{2} x$. Then the area of
$\triangle ABC$ is

$$\begin{aligned}
A &= \frac{1}{2}bh = \frac{1}{2}(BC)(AD) \\
&= \frac{1}{2}(x)\left(\frac{\sqrt{3}}{2}x\right) \\
&= \frac{\sqrt{3}}{4}x^2.
\end{aligned}$$

49. Transitive law

50. Subtraction property of inequalities

51. Division property of inequalities

52. Trichotomy law

53. The whole is greater than its parts

54. $\dfrac{x+3}{-2} > 7$

$-2\left(\dfrac{x+3}{-2}\right) < (-2)(7)$ Multiplication property of inequalities

$\qquad x+3 < -14$ Simplify

$\qquad x+3-3 < -14-3$ Subtraction property of inequalities

$\qquad\qquad x < -17$ Simplify

55. $2(y-1) < 3y+5$

$\qquad 2y-2 < 3y+5$ Distributive law

$2y-2+2 < 3y+5+2$ Addition property of inequalities

$\qquad 2y < 3y+7$ Simplify

$2y-3y < 3y-3y+7$ Subtraction property of inequalities

$\qquad -y < 7$ Simplify (Distributive law)

$(-1)(-y) > (-1)(7)$ Multiplication property of inequalities

$\qquad\qquad y > -7$

56. False; $m\angle B < m\angle 1$ since an exterior angle of a triangle is greater than each remote interior angle (Theorem 5.21).

57. True; an exterior angle of a triangle is greater than each remote interior angle (Theorem 5.21).

58. True, if two sides of a triangle are unequal, the angles opposite these sides are unequal in the same order (Theorem 5.22).

59. False; since $m\angle C = 53°$ and $m\angle 2 = 64°$, $m\angle B = 180° - (53° + 64°) = 63°$. Thus, $m\angle C < m\angle B < m\angle 2$ so by Theorem 5.23, $AB < AC < BC$.

60. True; by the Triangle Inequality, $AB < AC + BC = 11.5 + 10.8 = 22.3$ yd (Theorem 5.24)

61. $AC < DE$ by the SAS Inequality Theorem (Theorem 5.25)

62. No; since by the Triangle Inequality Theorem, each side must be less than the sum of the other two and $10 + 12 = 22 < 23$.

63. *Given:* $\triangle ABC$ with median $\overline{CP}$ and $m\angle APC > m\angle BPC$

Prove: $AC > BC$

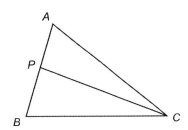

Proof: STATEMENTS	REASONS
1. ABC is a triangle with median $\overline{CP}$	1. Given
2. P is the midpoint of $\overline{AB}$	2. Def. of median
3. $AP = PB$	3. Def. of midpoint
4. $PC = PC$	4. Reflexive law
5. $m\angle APC > m\angle BPC$	5. Given
6. $AC > BC$	6. SAS Inequality Theorem

Chapter 5 Practice Test

1. $\dfrac{400 \text{ mi}}{10 \text{ hr}} = \dfrac{400 \text{ mi}}{10 \text{ hr}}$

$\phantom{\dfrac{400 \text{ mi}}{10 \text{ hr}}} = \dfrac{10 \cdot 40}{10} \dfrac{\text{mi}}{\text{hr}}$

$\phantom{\dfrac{400 \text{ mi}}{10 \text{ hr}}} = 40 \dfrac{\text{mi}}{\text{hr}}$

$\phantom{\dfrac{400 \text{ mi}}{10 \text{ hr}}} = 40 \text{ mph}$

2. $\dfrac{a}{a-2} = \dfrac{21}{15}$

$ 15a = 21(a-2)$

$ 15a = 21a - 42$

$ -6a = -42$

$ a = \dfrac{-42}{-6} = 7$

3. Let x be the geometric mean between 4 and 36. Then we must solve:

$$\frac{4}{x} = \frac{x}{36}$$

$$(4)(36) = x \cdot x$$

$$144 = x^2$$

Since $144 = (12)^2$ and $144 = (-12)^2$, $x = 12$ or $x = -12$.

4. Let x be the weight of 70 ft of wire. Then we must solve:

$$\frac{50}{65} = \frac{70}{x}$$

$$50x = (65)(70)$$

$$x = \frac{(65)(70)}{50} = 91$$

Thus, 70 ft of wire weighs 91 pounds.

5. True (This is the Means Property of Proportions)

6. Since $\square\; ABCD \sim \square\; A'B'C'D$,

$$\frac{AB}{A'B'} = \frac{BC}{B'C'}$$

$$\frac{18}{3} = \frac{12}{B'C'}$$

$$18B'C' = (3)(12)$$

$$B'C' = \frac{(3)(12)}{18} = 2$$

Thus, $B'C' = 2$ ft.

7. Yes

8. By Theorem 5.11,

$$\frac{x}{10} = \frac{6}{8}$$

$$8x = 60$$

$$x = 7.5$$

9. By Theorem 5.12,

$$\frac{y}{7} = \frac{BC}{AC}$$

Since $BC = x + 10$, from Problem 8, $x = 7.5$ so $BC = 7.5 + 10 = 17.5$. Also $AC = AD + DC = 6 + 8 = 14$. Thus,

$$\frac{y}{7} = \frac{17.5}{14}$$

$$14y = (7)(17.5)$$

$$y = \frac{(7)(17.5)}{14} = 8.75$$

10. Sometimes

11. Never

12. Since the triangles formed by the tower and its shadow and the building and its shadow are similar, if x is the height of the tower, we must solve:

$$\frac{x}{65} = \frac{160}{130}$$
$$130x = (65)(160)$$
$$x = \frac{(65)(160)}{130} = 80$$

Thus, the tower is 80 feet tall.

13.

STATEMENTS	REASONS
1. In given figure $\overline{AB} \parallel \overline{ED}$	1. Given
2. $\angle B \cong \angle D$; $\angle A \cong \angle E$	2. If lines $\parallel$, alt. int. $\angle$'s are $\cong$
3. $\triangle ABC \sim \triangle EDC$	3. AA

14. Let x be the geometric mean between 15 and 20. Then we must solve:

$$\frac{15}{x} = \frac{x}{20}$$
$$(15)(20) = (x)(x)$$
$$300 = x^2$$
$$\pm\sqrt{300} = x^2$$
$$\pm\sqrt{100 \cdot 3} = x$$
$$\pm 10\sqrt{3} = x$$

The solutions are $10\sqrt{3}$ and $-10\sqrt{3}$.

15. Since $\overline{XW}$ is a median of $\triangle XYZ$, W is the midpoint of $\overline{YZ}$, therefore $WZ = \frac{1}{2}(16) = 8$ inches. XW is $\frac{1}{2}(YZ)$ by Theorem 5.16. Thus, $XW = 8$ inches.

16. By Corollary 5.14,

$$\frac{AD}{BD} = \frac{BD}{CD}.$$

Substitute 5 for AD and 12 for CD.

$$\frac{5}{BD} = \frac{BD}{12}$$
$$(5)(12) = (BD)^2$$
$$60 = (BD)^2$$
$$\pm\sqrt{60} = BD$$
$$\pm\sqrt{4 \cdot 15} = BD$$
$$\pm 2\sqrt{15} = BD$$

Reject the negative value $-2\sqrt{15}$. Thus, $BD = 2\sqrt{15}$ cm.

17. By Corollary 5.15,

$$\frac{AC}{BC} = \frac{BC}{CD}.$$

Substitute 24 for AC and 18 for BC.

$$\frac{24}{18} = \frac{18}{CD}$$
$$24CD = (18)(18)$$
$$24CD = 324$$
$$CD = 13.5$$

Thus, $CD = 13.5$ yd.

18. Let $x = AD$. Then $CD = 20 - x$ since $AC = AD + DC$. By Corollary 5.14,

$$\frac{AD}{BD} = \frac{BD}{CD}.$$

Substitute x for AD, 8 for BD, and $20 - x$ for CD.

$$\frac{x}{8} = \frac{8}{20 - x}$$
$$x(20 - x) = (8)(8)$$
$$20x - x^2 = 64$$
$$0 = x^2 - 20x + 64$$
$$0 = (x - 4)(x - 16)$$
$$x - 4 = 0 \ \text{ or } \ x - 16 = 0$$
$$x = 4 \ \text{ or } \ \ \ \ \ \ x = 16$$

Thus, AD is either 4 ft or 16 ft.

19. If $m\angle 1 = 45°$, then $\angle C = 45°$ also making $\triangle BCD$ a $45°\text{-}45°\text{-}90°$ triangle. By Theorem 5.19, $BC = \sqrt{2}BD$. Thus, $6 = \sqrt{2}BD$. Divide both sides by $\sqrt{2}$.

$$BD = \frac{6}{\sqrt{2}} = \frac{6\sqrt{2}}{\sqrt{2}\,\sqrt{2}} = \frac{6\sqrt{2}}{2} = 3\sqrt{2} \ \text{cm.}$$

20. If $m\angle A = 60°$, then $m\angle ABD = 30°$ making $\triangle ABD$ a $30°\text{-}60°\text{-}90°$ triangle. By Corollary 5.20,

$$AD = \frac{1}{2}AB = \frac{1}{2}(14) = 7 \ \text{yd.}$$

21. The diagonal, d, is the hypotenuse of a right triangle with legs 12 yd. Use the

$$d^2 = 12^2 + 12^2$$
$$d^2 = 144 + 144$$
$$d^2 = 288$$
$$d = \pm\sqrt{288} = \pm\sqrt{144 \cdot 2} = \pm 12\sqrt{2}$$

Reject $-12\sqrt{2}$. Thus, the diagonal is $d = 12\sqrt{2} = 17.0$ yd, correct to the nearest tenth of a yard.

22. Use the Pythagorean Theorem.

$$a^2 + b^2 = c^2$$
$$(\sqrt{11})^2 + b^2 = 6^2$$
$$11 + b^2 = 36$$
$$b^2 = 25$$
$$b = \pm 5$$

Reject the value -5. Thus, $b = 5$ cm.

23. Consider the isosceles triangle with equal sides of length x and base angles $30°$, shown below.

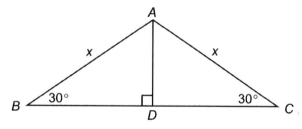

Construct altitude $\overline{AD}$ forming $30°\text{-}60°\text{-}90°$ triangle $\triangle ABD$. By Corollary 5.20, $AD = \frac{1}{2}x$ and $BD = \frac{\sqrt{3}}{2}x$. Then $BC = BD + DC = 2BD = \sqrt{3}x$. The area of $\triangle ABC$ is:

$$A = \frac{1}{2}bh = \frac{1}{2}(BC)(AD)$$
$$= \frac{1}{2}(\sqrt{3}x)\left(\frac{1}{2}x\right)$$
$$= \frac{\sqrt{3}}{4}x^2$$

24. Subtraction property of inequalities (The fact that $m\angle B > 30°$ makes $m\angle B - 10°$ positive.)

25.

$$4(x-3) > 5(x-2)$$

$4x - 12 > 5x - 10$	Distributive law
$4x - 12 + 12 > 5x - 10 + 12$	Addition property of inequalities
$4x > 5x + 2$	Simplify
$4x - 5x > 5x - 5x + 2$	Subtraction property of inequalities
$-x > 2$	Simplify (Distributive law)
$(-1)(-x) < (-2)(2)$	Multiplication property of inequalities
$x < -2$	Simplify

26. $m\angle E > m\angle B$ since an exterior angle of a triangle is greater than each remote interior angle.

27. $m\angle C < m\angle F$ by the SAS Inequality Theorem.

28. By the Triangle Inequality Theorem,

$AB < AC + BC$ so $10 < AC + 8$ or $2 < AC$
$AC < AB + BC$ so $AC < 10 + 8$ or $AC < 18$
$BC < AC + AB$ so $8 < AC + 10$ or $-2 < AC$

Thus, AC is less than 18 ft and more than 2 ft.

CHAPTER 6 CIRCLES

Section 6.1 Circles and Arcs

6.1 PRACTICE EXERCISE

1. Since $\angle ACB = 60°$ and $\angle ACB$ is an inscribed angle,

$$m\angle ACB = \frac{1}{2}m\overgroup{AB}$$

thus,

$$60° = \frac{1}{2}m\overgroup{AB}$$
$$120° = m\overgroup{AB}$$

Also,

$$m\overgroup{ACB} = 360° - m\overgroup{AB} = 360° - 120° = 240°.$$

6.1 SECTION EXERCISES

1. $d = 2r = 2(11) = 22$ in

3. $d = 2r = 2\left(\frac{3}{4}\right) = \frac{3}{2}$ ft

5. Since $d = 2r$, $r = \frac{d}{2} = \frac{16}{2} = 8$ in

7. Since $d = 2r$, $r = \frac{d}{2} = \frac{\frac{2}{3}}{2} = \frac{2}{3} \cdot \frac{1}{2} = \frac{1}{3}$ ft

9. a   **11.** b   **13.** d

15. h   **17.** j   **19.** e

21. Since the arcs are in the proportions 5: 6: 7,

$$5x + 6x + 7x = 360°$$
$$18x = 360°$$
$$x = 20$$

Thus $m\overgroup{XY} = 5(20) = 100°$; $m\overgroup{YZ} = 6(20) = 120$ and $m\overgroup{ZX} = 7(20) = 140°$.

23. To find $m\angle 4$, consider $\triangle YQZ$. It is isosceles since 2 sides are radii, thus the base angles are = in measure. The sum of the $\angle$'s $= 180°$ thus if $x = m\angle 4$,

$$x + x + 120 = 180$$
$$2x = 60$$
$$x = 30 \text{ thus } m\angle 4 = 30°$$

25. **(a)** central angle   **(b)** inscribed angle

(c) Since central angle $m\angle AOC = 120°$,

$$m\overgroup{AC} = 120°$$

(d) $m\overgroup{ABC} = 360° - m\overgroup{AC}$
$$= 360° - 120° = 240°$$

(e) $m\angle ABC = \frac{1}{2}m\overgroup{AC} = \frac{1}{2}(120°) = 60°$

(f) The arc intercepted by $\angle ABC$ is twice the measure of $\angle ABC$, which is $2(60°) = 120°$.

27. **(a)** Since $\angle CBA$ is an inscribed angle $m\angle CBA = 32°$, the measure of central angle $\angle COA = 64°$. Thus, $m\overgroup{ADC} = 64°$ also.

(b) $m\overgroup{ABC} = 360° - 64° = 296°$.

(c) $m\angle AOC = 64°$ (twice inscribed angle $\angle ABC$ which is $32°$).

(d) $m\angle AEC = m\angle ABC = 32°$.

29. *Proof:*

STATEMENTS	REASONS
1. $\overline{AB} \perp \overline{BC}$	1. Given
2. $\angle ABC$ is a right angle	2. $\perp$ lines form rt. $\angle$'s
3. $m\angle ABC = 90°$	3. Def. of rt. $\angle$
4. $\frac{1}{2}m\overset{\frown}{ADC} = m\angle ABC$	4. Measure of inscribed angle = $\frac{1}{2}$ measure intercepted arc.
5. $m\overset{\frown}{ADC} = 2m\angle ABC$	5. Mult.-Div. Prop.
6. $m\overset{\frown}{ADC} = 2(90°) = 180°$	6. Substitution law

31. *Construction:* Draw radius $\overline{OC}$

Proof:

STATEMENTS	REASONS
1. $\overline{AB}$ is a diameter	1. Given
2. $\overline{AC} \parallel \overline{OD}$	2. Given
3. $\overline{OC}$ is a radius	3. By construction
4. $\angle CAO \cong \angle DOB$; $m\angle CAO = m\angle DOB$	4. If $\parallel$ lines, corr. $\angle$'s are $\cong$; Def. $\cong \angle$'s
5. $\overline{AO} \cong \overline{OC}$	5. Radii are $\cong$
6. $\angle CAO \cong \angle ACO$	6. $\angle$'s opp. $\cong$ sides of $\triangle$ are $\cong$
7. $\angle ACO \cong \angle COD$	7. If $\parallel$ lines, alt. int. $\angle$'s are $\cong$
8. $\angle CAO \cong \angle COD$; $m\angle CAO = m\angle COD$	8. Transitive law; Def. $\cong \angle$'s
9. $m\angle DOB = m\angle COD$	9. Substitution law
10. $m\overset{\frown}{BD} = m\angle DOB$ and $m\overset{\frown}{DC} = m\angle COD$	10. Def. of measure of arc
11. $m\overset{\frown}{BD} = m\overset{\frown}{DC}$	11. Trans. law

33. Yes, by the definition of a semicircle.

Section 6.2 Chords and Secants

6.2 PRACTICE EXERCISES

1. Since $m\widehat{AD} = 112°$ and $m\widehat{BC} = 176°$, by Theorem 6.5,

$$m\angle APD = \frac{1}{2}(m\widehat{AD} + m\widehat{BC})$$

$$= \frac{1}{2}(112° + 176°)$$

$$= \frac{1}{2}(288°)$$

$$= 144°$$

Since $\angle 1$ and $\angle APD$ are supplementary,

$$m\angle 1 + m\angle APD = 180°$$
$$m\angle 1 + 144° = 180°$$
$$m\angle 1 = 180° - 144°$$
$$m\angle 1 = 36°$$

2. The reason for Statement 1 is: Given
Statement 2 is: $\overline{AO} \cong \overline{BO}$
Both are radii and radii are equal.
The reason for Statement 3 is: Def. of isosc. $\triangle$.

Statement 4 is: $\overline{AB}$ is the base of $\triangle AOB$
Statement 5 is: $\overline{CE}$ passes through 0.

3. By Theorem 6.13,

$$(AP)(PB) = (CP)(PD).$$

Substituting we have:

$$3(PB) = (4)(6)$$
$$3PB = 24$$
$$PB = 8 \text{ cm}$$

6.2 SECTION EXERCISES

1. $m\angle AEB = \frac{1}{2}(m\widehat{AB} + m\widehat{CD})$

$$= \frac{1}{2}(58° + 130°)$$

$$= \frac{1}{2}(188°) = 94°$$

3. Since equal chords form equal arcs, $m\widehat{AB} = 60°$.

5. Since a line through the center of a circle perpendicular to a chord bisects the chord, $AE = CE$. Thus, $CE = 12$ in.

7. Since $\overline{BD}$ passes through the center of the circle and bisects AC (we are given $AE = CE = 5$ ft), it is perpendicular to chord $\overline{AC}$ making $m\angle AED = 90°$.

9. By Theorem 6.13, $(AE)(EC) = (BE)(ED)$. Substituting 5 for AE, 6 for EC, and 3 for BE we have $(5)(6) = (3)(ED)$ which gives $30 = 3ED$ or $ED = 10$ cm.

11. By Theorem 6.10, since $AB = CD = 14$ ft, $OE = OF$ so that $OE = 11$ ft.

13. Since $OE = OF = 5$ cm, by Theorem 6.11, $AB = CD$. Since $\overline{OE} \perp \overline{AB}$ and $\overline{OF} \perp \overline{CD}$, $\overline{OE}$ bisects $\overline{AB}$ and $\overline{OF}$ bisects $\overline{CD}$ by Theorem 6.8. Thus $AE = DF$, and since $AE = 4$ cm, $DF = 4$ cm.

15. Since $m\widehat{AGB} = m\widehat{DHC}$, by Theorem 6.7, $AB = DC$. Since $AB = DC$, by Theorem 6.10, $OE = OF$. Since $OE = 15$ cm, $OF = 15$ cm.

17. $m\angle P = \frac{1}{2}(m\overset{\frown}{AB} - m\overset{\frown}{CD})$

$\qquad = \frac{1}{2}(130° - 50°)$

$\qquad = \frac{1}{2}(80°) = 40°$

19. $m\angle P = \frac{1}{2}(m\overset{\frown}{AB} - m\overset{\frown}{CD}$, substitute 120° for

$\overset{\frown}{AB}$ and 45° for $m\angle P$ and solve for $m\overset{\frown}{CD}$.

Let $x = m\overset{\frown}{CD}$

$\quad 45° = \frac{1}{2}(120° - x)$

$\quad 90° = 120° - x$ Multiply both
$\qquad\qquad\qquad\qquad$ sides by 2

$\quad -30° = -x$ $\qquad$ Subtract 120°
$\qquad\qquad\qquad\qquad$ from both sides

$\quad 30° = x$ $\qquad\quad$ Multiply both
$\qquad\qquad\qquad\qquad$ sides by −1

Thus $m\overset{\frown}{CD} = 30°$

21. By Theorem 6.15, $(AP)(CP) = (PC)(PD)$.
Substitute 12 for AP, 5 for CP, and 15 for PB.

$\qquad (12)(5) = (15)(PD)$

$\qquad \frac{(12)(5)}{15} = PD$

$\qquad\qquad 4 = PD$

Thus, $PD = 4$ ft.

23. By Theorem 6.15, $(AP)(CP) = (BP)(DP)$.
Since $AP = AC + CP = 5 + 4 = 9$ cm, and
$BP = BD + DP = BD + 3$, substituting we
have:

$\qquad (9)(4) = (BD + 3)(3)$

$\qquad\quad 36 = 3BD + 9$

$\qquad\quad 27 = 3BD$

$\qquad\quad\ 9 = BD$

Thus, $BD = 9$ cm.

25. Since $m\angle DAB = 30°$, $m\overset{\frown}{BD} = 60°$. By

Theorem 6.5, $m\angle 1 = \frac{1}{2}(\overset{\frown}{BD} + \overset{\frown}{AC}$ so

$\quad 40° = \frac{1}{2}(60° + m\overset{\frown}{AC})$.

Then $80° = 60° + m\overset{\frown}{AC}$ so that $m\overset{\frown}{AC} = 20°$.
Since

$m\angle ABC = \frac{1}{2}m\overset{\frown}{AC}, m\angle ABC = \frac{1}{2}(20°) = 10°$.

27. *Proof:*

STATEMENTS	REASONS
1. $m\overset{\frown}{BC} + m\overset{\frown}{AD} = 180°$	1. Given
2. $m\angle AED = 90°$	2. $\angle$ formed by chords is $\frac{1}{2}$ sum arc + arc of vertical $\angle$ (Theorem 6.5)
3. $m\overset{\frown}{AB} + m\overset{\frown}{CD} + m\overset{\frown}{BC} + m\overset{\frown}{AD} = 360°$	3. A circle is an arc of 360°.
4. $m\overset{\frown}{AB} + m\overset{\frown}{CD} = 180°$	4. Add.-Subt. Post.
5. $m\angle AED = 90°$	5. Same as 2
6. $\overline{AC} \perp \overline{BD}$	6. Def. of $\perp$ lines

29. *Proof:*   STATEMENTS   REASONS

    1. $\overline{AB} \cong \overline{BC}$

    1. Given

    2. $m\angle P = \dfrac{1}{2}(m\overset{\frown}{BC} - m\overset{\frown}{AD})$

    2. If secants intersect outside $\odot$, measure of $\angle = \dfrac{1}{2}$ difference of intercepted arcs. (Thm 6.14).

    3. $\overset{\frown}{AB} \cong \overset{\frown}{BC}$ thus $m\overset{\frown}{AB} = m\overset{\frown}{BC}$

    3. $\cong$ chords have $\cong$ arcs; Def. $\cong$ arc.

    4. $m\angle P = \dfrac{1}{2}(m\overset{\frown}{AB} - m\overset{\frown}{AB})$

    4. Substitution law

31. *Given:* $\overset{\frown}{AB} \cong \overset{\frown}{CD}$

Prove: $\overline{AB} \cong \overline{CD}$

Construction: Draw radii $\overline{AO}$, $\overline{BO}$, $\overline{CO}$, and $\overline{DO}$

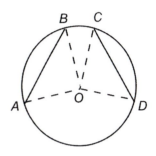

Proof: STATEMENTS   REASONS

    1. $\overset{\frown}{AB} \cong \overset{\frown}{CD}$

    1. Given

    2. $\overline{AO}$, $\overline{BO}$, $\overline{CO}$, and $\overline{DO}$ are radii

    2. By construction

    3. $\overline{AO} \cong \overline{BO} \cong \overline{CO} \cong \overline{DO}$

    3. Radii are $\cong$

    4. $m\angle BOA = m\angle COD$ thus $\angle BOA \cong \angle COD$

    4. Def. of measure of an arc; Def. $\cong \angle$'s

    5. $\triangle BOA \cong \triangle COD$

    5. SAS

    6. $AB = CD$

    6. CPCTC

33. *Given:* $\overline{AB}$ and $\overline{CD}$ are chords with $\overline{OE} \perp \overline{AB}$, $\overline{OF} \perp \overline{CD}$, and $OE = OF$

Prove: $\overline{AB} \cong \overline{CD}$

Construction: Draw radii $\overline{AO}$, $\overline{BO}$, $\overline{CO}$, and $\overline{DO}$

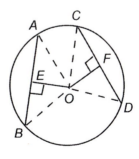

Proof: STATEMENTS   REASONS

1. $\overline{AB}$ and $\overline{CD}$ are chords

1. Given

2. $\overline{OE} \perp \overline{AB}$ and $\overline{OF} \perp \overline{CD}$

2. Given

3. $\angle AEO$, $\angle BEO$, $\angle CFO$, and $\angle DFO$ are right angles

3. $\perp$ lines form rt. $\angle$'s

4. $OE = OF$

4. Given

5. $\overline{AO} \cong \overline{BO} \cong \overline{CO} \cong \overline{DO}$

5. Radii are $\cong$

6. $\triangle AEO$, $\triangle BEO$, $\triangle CFO$, and $\triangle DFO$ are right triangles

6. Def. of rt. $\triangle$

7. $\triangle AEO \cong \triangle CFO$ and $\triangle BEO \cong \triangle DFO$

7. HL

8. $\overline{AE} \cong \overline{CF}$ and $\overline{BE} \cong \overline{DF}$ thus $AE = CF$ and $BE = DF$

8. CPCTC; Def. $\cong$ seg.

9. $AE + BE = CF + DF$

9. Add.-Subt. Post.

10. $AB = AE + BE$ and $CD = CF + DF$

10. Seg.-Add. Post.

11. $AB = CD$ thus $\overline{AB} \cong \overline{CD}$

11. Substitution law; Def. $\cong$ seg.

35. Suppose $AB = 10$ inches, P is the midpoint of $\overarc{AB}$, $\overline{PQ} \perp \overline{AB}$, and $PQ = 1$ inch as shown in the figure to the right. We want to find PO. Extend PO to form segment PR. By Theorem 6.13, $(PQ)(QR) = (AQ)(BQ)$. Substitute 5 for AQ and BQ (a $\perp$ line bisects the chord) and 1 for PQ.

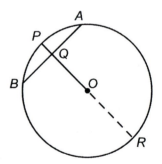

$$(1)(QR) = (5)(5)$$
$$QR = 25$$

Then $PQ + QR = 1 + 25 = 26$ is the length of a diameter making the radius $\dfrac{1}{2}(26) = 13$ inches.

Section 6.3 Tangents

6.3 PRACTICE EXERCISES

1. Using Figure 6.25, as in Example 1, we have

$$(OS)^2 = (SA)^2 + (OA)^2$$
$$(SA)^2 = (OS)^2 - (OA)^2$$
$$= [OB + SB]^2 - (OA)^2$$
$$= [4000 + 95]^2 - (4000)^2$$
$$= [4095]^2 - (4000)^2$$
$$= 769,025$$

Talking the square root with a calculator we obtain
$$SA = 876.9407$$

or, to the nearest mile, the horizon is about 877 miles away.
Only the positive square root is used because SA is a distance.

2. By Theorem 6.18,

$$m\angle APD = \frac{1}{2}(m\widehat{ACD} - m\widehat{ABD})$$

Since $m\widehat{ACD} = 236°$, and
$$m\widehat{ACD} + m\widehat{ABD} = 360°,$$
$$m\widehat{ABD} = 360° - 236° = 124°. \text{ Substituting}$$
we have:

$$m\angle APD = \frac{1}{2}(236° - 124°)$$
$$= \frac{1}{2}(112°)$$
$$= 56°$$

6.3 SECTION EXERCISES

1. By Theorem 6.16,

$$m\angle PAD = \frac{1}{2}m\widehat{AD} = \frac{1}{2}(70°) = 35°$$

3. Since $m\angle 1 = 36°$, by Theorem 6.2,

$$m\angle 1 = \frac{1}{2}m\widehat{AD}. \text{ Thus, } 36° = \frac{1}{2}m\widehat{AD} \text{ making}$$

$$m\widehat{AD} = 72°.$$
By Theorem 6.16,
$$m\angle PAD = \frac{1}{2}m\widehat{AD} = \frac{1}{2}(72°) = 36°.$$

5. By Theorem 6.16, $m\angle PAD = \frac{1}{2}m\widehat{AD}$ and

by Theorem 6.2, $m\angle 1 = \frac{1}{2}m\widehat{AD}.$

Thus, $m\angle 1 = m\angle PAD = 40°.$

7. By Theorem 6.17,

$$m\angle APD = \frac{1}{2}(m\widehat{AB} - m\widehat{AD})$$
$$= \frac{1}{2}(140° - 70°)$$
$$= \frac{1}{2}(70°) = 35°.$$

9. By Theorem 6.2, $m\angle 1 = \frac{1}{2}m\widehat{AD}$ and
$m\angle 3 = \frac{1}{2}m\widehat{AB}.$ Then $36° = \frac{1}{2}m\widehat{AD}$ and
$70° = \frac{1}{2}m\widehat{AB}$ so that $m\widehat{AD} = 72°$ and
$m\widehat{AB} = 140°.$ By Theorem 6.7,

$$m\angle APD = \frac{1}{2}(m\widehat{AB} - m\widehat{AD})$$
$$= \frac{1}{2}(140° - 72°)$$
$$= \frac{1}{2}(68°) = 34°.$$

11. By Theorem 6.17,

$$m\angle APD = \frac{1}{2}(m\overset{\frown}{AB} - m\overset{\frown}{AD})$$

Thus $40° = \frac{1}{2}(138° - m\overset{\frown}{AD})$ so that

$80° = 138° - m\overset{\frown}{AD}$ which means $m\overset{\frown}{AD} = 58°$.
By Theorem 6.2,

$$m\angle 1 = \frac{1}{2}m\overset{\frown}{AD} = \frac{1}{2}(58°) = 29°.$$

13. Since $m\overset{\frown}{ADC} = 130°$ and
$m\overset{\frown}{ABC} = 360° - m\overset{\frown}{ADC}$,
$m\overset{\frown}{ABC} = 360° - 130° = 230°$.

By Theorem 6.18,

$$m\angle APC = \frac{1}{2}(m\overset{\frown}{ABC} - m\overset{\frown}{ADC})$$

$$= \frac{1}{2}(230° - 130°)$$

$$= \frac{1}{2}(100°) = 50°.$$

15. Since $m\angle 1 = 35°$ and $m\angle 2 = 30°$, by
Theorem 6.2, $m\overset{\frown}{AD} = 70°$ and $m\overset{\frown}{CD} = 60°$.
Then
$m\overset{\frown}{ADC} = m\overset{\frown}{AD} + m\overset{\frown}{CD} = 70° + 60° = 130°$,
making $m\overset{\frown}{ABC} = 360° - m\overset{\frown}{ABC} = 360° - 130°$
$= 230°$. By Theorem 6.18,

$$m\angle APC = \frac{1}{2}(m\overset{\frown}{ABC} - m\overset{\frown}{ADC})$$

$$= \frac{1}{2}(230° - 130°)$$

$$= \frac{1}{2}(100°) = 50°$$

17. By Theorem 6.19, $AP = CP$ so that
$CP = 17$ cm.

19. By Theorem 6.20, AP is the geometric mean
between BP and DP. Thus,

$$\frac{BP}{AP} = \frac{AP}{DP}$$

$$\frac{18}{12} = \frac{12}{DP}$$

$$18DP = (12)(12)$$

$$DP = \frac{(12)(12)}{18} = 8 \text{ cm}$$

21. Let $BD = x$, then $BP = x + DP = x + 10$.
By Theorem 6.20, AP is the geometric mean
between BP and DP. Thus,

$$\frac{BP}{AP} = \frac{AP}{DP}$$

$$\frac{x+10}{15} = \frac{15}{10}$$

$$10(x+10) = (15)(15)$$

$$10x + 100 = 225$$

$$10x = 125$$

$$x = 12.5$$

Thus, $BD = 12.5$ yd.

23. There are two internal tangents, m and n, to
circles in the given positions as shown in the
figure below.

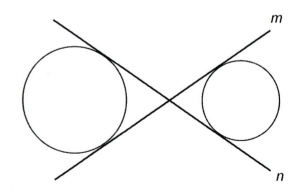

25. *Proof:*

STATEMENTS	REASONS
1. $\overleftrightarrow{PA}$ and $\overleftrightarrow{PB}$ are tangents to the circle	1. Given
2. $m\angle P = \dfrac{1}{2}(m\overset{\frown}{ADB} - m\overset{\frown}{ABC})$	2. Theorem 6.18
3. $m\overset{\frown}{ADB} + m\overset{\frown}{ACB} = 360°$	3. Circle measures $360°$
4. $m\overset{\frown}{ADB} = 360° - m\overset{\frown}{ACB}$	4. Arc Add. Post.
5. $m\angle P = \dfrac{1}{2}([360° - m\overset{\frown}{ACB}] - m\overset{\frown}{ACB})$	5. Substitution law
6. $m\angle P = \dfrac{1}{2}(360° - 2m\overset{\frown}{ACB})$	6. Distributive law
7. $m\angle P = 180° - m\overset{\frown}{ACB}$	7. Distributive law
8. $m\angle P + m\overset{\frown}{ACB} = 180°$	8. Add.-Subt. Post.

27. *Construction:* Extend $\overleftrightarrow{AB}$ and $\overleftrightarrow{CD}$ to meet at point P

Proof:

STATEMENTS	REASONS
1. $\overleftrightarrow{AB}$ and $\overleftrightarrow{CD}$ are common external tangents to circles that are not congruent.	1. Given
2. $\overleftrightarrow{AB}$ and $\overleftrightarrow{CD}$ meet at P	2. By Construction since circles are not $\cong$
3. $AP = CP$ and $BP = DP$	3. Theorem 6.19
4. $AB = AP - BP$ and $CD = CP - DP$	4. Seg. Add. Post.
5. $AP - BP = CP - DP$	5. Add.-Subt. Post.
6. $AB = CD$	6. Substitution law

29. *Given:* $\overleftrightarrow{AP}$ and $\overleftrightarrow{BP}$ are tangents

Prove: $m\angle APB = \dfrac{1}{2}(m\overset{\frown}{ADB} - m\overset{\frown}{ACB})$

Construction: Choose point D on major arc AB and draw secant $\overrightarrow{PD}$ intersecting the circle at C and D.

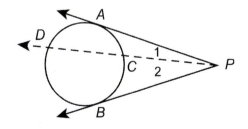

Proof: STATEMENTS   REASONS

1. $\overleftrightarrow{AP}$ and $\overleftrightarrow{BP}$ are tangents   1. Given

2. $\overrightarrow{PD}$ is a secant   2. By construction

3. $m\angle 1 = \frac{1}{2}(m\widehat{AD} - m\widehat{AC})$ and $m\angle 2 = \frac{1}{2}(m\widehat{DB} - m\widehat{BC})$ 3. Theorem 6.17

4. $m\angle APB = m\angle 1 + m\angle 2$ 4. $\angle$ Add. Post.

5. $m\angle 1 + m\angle 2 = \frac{1}{2}(m\widehat{AD} - m\widehat{AC}) + \frac{1}{2}(m\widehat{DB} - m\widehat{BC})$ 5. Add.-Subt. Post.

6. $m\angle 1 + m\angle 2 = \frac{1}{2}[m\widehat{AD} + m\widehat{DB} - (m\widehat{AC} + m\widehat{BC})]$ 6. Simplify

7. $m\widehat{AD} + m\widehat{DB} = m\widehat{ADB}$ and $m\widehat{AC} + m\widehat{BC} = m\widehat{ACB}$ 7. Arc Add. Post.

8. $m\angle APB = \frac{1}{2}(m\widehat{ADB} - m\widehat{ACB})$ 8. Substitution law

31. Use Construction 6.1.   **33.** Use Construction 6.3.   **35.** Use Construction 6.4

37. Follow the solution to Example 1: Using same notation and figure we have:

$$(SA)^2 = (OS)^2 - (OA)^2$$
$$= [OB + SB]^2 - (OA)^2$$
$$= [4000 + 1]^2 - (4000)^2$$
$$= (4001)^2 - (4000)^2$$
$$= 8001$$

Taking the square root of both sides we have $SA = 89.4$ miles, correct to the nearest tenth of a mile.

39.

Section 6.4 Circles and Regular Polygons

6.4 PRACTICE EXERCISES

1. First divide the circle into 8 equal arcs using the same method as was shown in Example 1. Then use Construction 6.1 to construct the tangent to the circle at each of these eight points on the circle. The points of intersection of these tangents form the vertices of the desired circumscribed octagon.

2. Since the measure of each angle of a regular n-gon is given by

$$a = \frac{360°}{n},$$

if $a = 24$, substitute and solve for n.

$$24° = \frac{360°}{n}$$
$$(24°)n = 360°$$
$$n = \frac{360°}{24°} = 15$$

Thus, the regular polygon has 15 sides.

6.4 SECTION EXERCISES

1. By Theorem 6.23, $\angle A$ and $\angle C$ are supplementary. Thus $m\angle A + m\angle C = 180°$; $86° + m\angle C = 180°$; $m\angle C = 180° - 86° = 94°$.

3. Since $\angle A$ and $\angle C$ are supplementary (by Theorem 6.23), $m\angle A + m\angle C = 180°$. Substituting into $m\angle A + m\angle B + m\angle C = 276°$ we have $180° + m\angle B = 276°$ making $m\angle B = 276° - 180° = 96°$.

5. With $\overline{AB} \parallel \overline{DC}$ and $\overline{AB} \cong \overline{DC}$, opposite sides of the quadrilateral are parallel and $\cong$ making the quadrilateral a rectangle. Since all angles are 90°, $m\angle A = 90°$.

7. Since $\angle A \cong \angle C$, $m\angle A = m\angle C$ and $m\angle A + m\angle C = 180°$ (by Theorem 6.23), $2m\angle C = 180°$ making $m\angle C = 90°$.

9. The endpoints of a diameter and the points of intersection of the circle with the perpendicular bisector of the diameter give four points that divide the circle into four equal arcs. By Theorem 6.25, the chords formed by these arcs form a regular 4-gon, that is, form a square.

11. (g) Yes, the polygon is equilateral by Theorem 6.7. The sides of the polygon are chords of the circle. Chords formed by $\cong$ arcs are $\cong$.

(h) Yes, the polygon is equiangular. First construct all radii of the polygon. They are $\cong$ to each other and the chords of the circle thus forming six equilateral $\triangle$s. Equilateral $\triangle$s are equiangular. Thus the polygon is equiangular.

13. Since the length of a side of a regular inscribed hexagon is equal to the radius, by marking off six points on the circle with a compass set at the length of the radius, and joining every other point we form an equilateral triangle that is inscribed in the circle (by Theorem 6.25).

15. Use Construction 4.2 and Construction 6.6.

17. Use Construction 2.4 and Construction 6.5.

19. By Theorem 6.29, each central angle measures $\dfrac{360°}{n}$ where $n = 3$. Thus, each angle is $\dfrac{360°}{3} = 120°$.

21. By Theorem 6.29, each central angle measures $\dfrac{360°}{n}$ where $n = 5$. Thus each angle is $\dfrac{360°}{5} = 72°$.

23. By Theorem 6.29, each central angle measures $\dfrac{360°}{n}$ where $n = 9$. Thus each angle is $\dfrac{360°}{9} = 40°$.

25. By Theorem 6.29, $45° = \dfrac{360°}{n}$. Thus,, $45n = 360$ so $n = \dfrac{360}{45} = 8$. The polygon has 8 sides (an octagon).

27. By Theorem 6.29, $20° = \dfrac{360°}{n}$. Thus, $20n = 360$ so $n = \dfrac{360}{20} = 18$. The polygon has 18 sides.

29. *Given:* ABCD is a parallelogram inscribed in a circle

Prove: ABCD is a rectangle

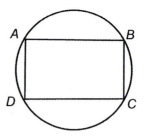

Proof: STATEMENTS

1. ABCD is a parallelogram
2. $\overline{AB} \cong \overline{DC}$ and $\overline{AD} \cong \overline{BC}$
3. $\overset{\frown}{AB} \cong \overset{\frown}{DC}$ and $\overset{\frown}{AD} \cong \overset{\frown}{BC}$ thus $m\overset{\frown}{AB} = m\overset{\frown}{DC}$ and $m\overset{\frown}{AD} = m\overset{\frown}{BC}$
4. $m\overset{\frown}{AB} + m\overset{\frown}{AD} = m\overset{\frown}{DC} + m\overset{\frown}{BC}$
5. $m\overset{\frown}{BAD} = m\overset{\frown}{AB} + m\overset{\frown}{AD}$ and $m\overset{\frown}{BCD} = m\overset{\frown}{DC} + m\overset{\frown}{BC}$
6. $m\overset{\frown}{BAD} = m\overset{\frown}{BCD}$
7. $m\angle A = \frac{1}{2} m\overset{\frown}{BAD}$ and $m\angle C = \frac{1}{2} m\overset{\frown}{BCD}$
8. $m\angle A = m\angle C$
9. $\angle A$ and $\angle C$ are supplementary
10. $m\angle A + m\angle C = 180°$
11. $m\angle A + m\angle A = 180°$
12. $2m\angle A = 180°$
13. $m\angle A = 90°$
14. $\angle A$ is a right angle
15. ABCD is a rectangle

REASONS

1. Given
2. Opp. sides of $\square$ are $\cong$
3. $\cong$ chords from $\cong$ arcs; Def. $\cong$ arcs
4. Add.-Subt. Post.
5. Arc Add. Post.

6. Substitution law
7. Measure inscribed $\angle = \frac{1}{2}$ measure intercepted arc

8. Substitution
9. Opp. $\angle$'s of inscribed quad. are supp.
10. Def. of supp. $\angle$'s
11. Substitution law
12. Distributive law
13. Mult.-Div. Post.
14. Def. of rt. $\angle$
15. Def. of $\square$

31. The proof follows from the fact that each side is an equal chord of the <u>circumscribed</u> circle; hence the same distance from the center of the circle so that all radii of the inscribed circle must be equal in length.

33. The proof follows by circumscribing a circle around the regular polygon and using the fact that $\cong$ chords (the equal sides of the regular polygon) have $\cong$ central angles.

35. Consider an equilateral triangle inscribed in a circle with radius 4 cm as shown to the right. Draw altitude $\overline{OP}$ to side $\overline{AB}$. In right triangle $\triangle PBO$, $m\angle PBO = 30°$, $m\angle POB = 60°$, and hypotenuse $\overline{BO}$ (radius $\overline{BO}$) = 4 cm. Thus, by Theorem 5.20,

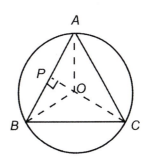

$$PB = \frac{\sqrt{3}}{2} BO = \frac{\sqrt{3}}{2}(4) = 2\sqrt{3}. \text{ Since } AB =$$

$2PB$, $AB = 2(2\sqrt{3}) = 4\sqrt{3}$ cm.

Section 6.5 Inequalities Involving Circles (optional)

6.5 PRACTICE EXERCISE

1. The reason for Statement 1 is: Given

   Statement 2 is: $m\angle 1 > m\angle 2$

   Note that $m\overarc{AB} > m\overarc{CD}$ and $m\angle 1 = \overarc{AB}$ and $m\angle 2 = \overarc{CD}$ so $m\angle 1 > m\angle 2$

   Statement 3 is: $\overline{OA} \cong \overline{OB} \cong \overline{OC} \cong \overline{OD}$

   All radii are $\cong$.

   The reason for Statement 4 is: SAS Ineq. Thm.

6.5 SECTION EXERCISES

1. True; the greater of two central angles intercepts the greater of the two arcs (Theorem 6.31).

3. True; the greater of two unequal chords is nearer the center of a circle (Theorem 6.34).

5. False; since $m\overarc{AB} > m\overarc{CD}$, $AB > CD$ so $\overline{AB}$ is closer to the center of the circle than $\overline{CD}$ by Theorems 6.33 and 6.34.

7. True; a 60° angle determines a chord equal in length to a radius so the chord determined by a 45° angle is shorter than the chord determined by a 60° angle.

9. $m\overarc{AB} > m\overarc{CD}$ since central angle $\angle 1$ is greater than central angle $\angle 2$ (Theorem 6.30).

11. $m\overarc{AB} > m\overarc{CD}$ since $AB = 13.5 > 12.5 = CD$ and the greater of two chords forms the arc (Theorem 6.32).

13. $AB > CD$ since $m\overarc{AB} = 125° > 120° = m\overarc{CD}$ and the greater of two arcs has the greater chord (Theorem 6.33).

15. $\overline{CD}$ is farther from the center since $CD = 9.8 < 10.2 = AB$ and the greater of two unequal chords is nearer the center of the circle (Theorem 6.34).

17. *Given:* Congruent circles with centers
   O and P, $m\angle 1 > m\angle 2$

Prove: $m\overset{\frown}{AB} > m\overset{\frown}{CD}$

Construction: Construct $\overset{\frown}{ABD} \cong \overset{\frown}{AB}$ and
   draw chords $\overline{AB}$ and $\overline{CE}$

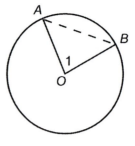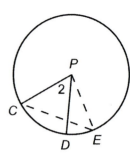

Proof:

STATEMENTS	REASONS
1. The circles centered at O and P are congruent	1. Given
2. $\overline{AO}$, $\overline{BO}$, $\overline{CP}$, and $\overline{EP}$ are radii	2. Def. of radii
3. $\overline{AO} \cong \overline{BO} \cong \overline{CP} \cong \overline{EP}$	3. Radii of same and $\cong$ circles are $\cong$
4. $\overset{\frown}{ABD} \cong \overset{\frown}{AB}$	4. By construction
5. $\overline{AB} \cong \overline{CE}$	5. Chords formed by $\cong$ arcs are $\cong$ (Thm 6.7)
6. $\triangle AOB \cong \triangle CPE$	6. SSS
7. $\overline{AB} \cong \overline{CE}$; $\angle 1 \cong \angle CPE$	7. CPCTC
8. $m\overset{\frown}{AB} = m\overset{\frown}{CE}$	8. Def. of arc measure
9. $m\angle 1 > m\angle 2$	9. Def. of arc measure
10. $m\angle CPE > m\angle 2$	10. Given
11. $m\overset{\frown}{CE} > m\overset{\frown}{CD}$	11. Substitution law
12. $m\overset{\frown}{AB} > m\overset{\frown}{CD}$	12. In same circle, > of 2 central $\angle$'s has > arc (Thm 6.30)

19. *Given:* Congruent circles with centers
   O and P, $m\overset{\frown}{AB} > m\overset{\frown}{CD}$

Prove: $m\angle 1 > m\angle 2$

Construction: Construct $\overset{\frown}{CDE} \cong \overset{\frown}{AB}$ and
   draw chords $\overline{AB}$, $\overline{CD}$, and $\overline{CE}$.

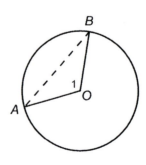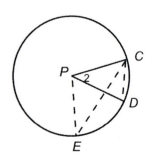

Proof: STATEMENTS	REASONS
1. The circles with centers O and P congruent	1. Given
2. $\overline{AO}$, $\overline{BO}$, $\overline{CP}$, $\overline{DP}$, and $\overline{EP}$ are radii	2. Def. of radii
3. $\overline{AO} \cong \overline{BO} \cong \overline{CP} \cong \overline{DP} \cong \overline{EP}$	3. Radii of same and $\cong$ circles are $\cong$
4. $\overparen{CDE} \cong \overparen{AB}$	4. By Construction
5. $\overline{AB} \cong \overline{CE}$	5. Chords formed by congruent arcs congruent
6. $\triangle AOB \cong \triangle CPE$	6. SSS
7. $\overline{AB} \cong \overline{CE}$	7. CPCTC
8. $\overline{AB} \cong \overline{CE}$ thus $m\overparen{AB} = m\overparen{CE}$	8. $\cong$ chords have $\cong$ arcs; Def. $\cong$ arcs
9. $m\overparen{AB} > m\overparen{CD}$	9. Given
10. $m\overparen{CE} > m\overparen{CD}$	10. Substitution
11. $m\angle CPE > m\angle 2$	11. Exercise 18
12. $m\angle 1 > m\angle 2$	12. Substitution law

21. *Given:* Congruent circles O and P,
arcs $\overparen{AB}$ and $\overparen{CD}$ with $m\overparen{AB} > m\overparen{CD}$

Prove: $AB > CD$

Construction: Draw radii $\overline{AO}$, $\overline{BO}$, $\overline{CP}$, and $\overline{DP}$ and chords $\overline{AB}$ and $\overline{CD}$

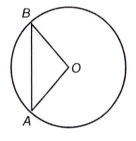 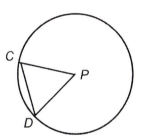

Proof: STATEMENTS	REASONS
1. Congruent circles O and P with arcs $\overparen{AB}$ and $\overparen{CD}$	1. Given
2. $m\overparen{AB} > m\overparen{CD}$	2. Given
3. $m\overparen{AB} = m\angle AOB$ and $m\overparen{CD} = m\angle CPD$	3. Def. of arc measure
4. $m\angle AOB > m\angle CPD$	4. Substitution law
5. $\overline{AO} \cong \overline{BO} \cong \overline{CP} \cong \overline{DP}$	5. Radii of same or $\cong$ circles are $\cong$
6. $AB > CD$	6. SAS Inequality Theorem

23. The proof is in two parts: The case for the same circle first, then the case for congruent circles. We present the first case, the second follows from it after using a technique similar to that given in the proof of Theorem 8.11.

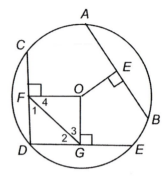

Given: The circle centered at O, chords $\overline{AB}$ and $\overline{CD}$ with $\overline{OE} \perp \overline{AB}$, $\overline{OF} \perp CD$ and $OF > OE$.

Prove: $AB > CD$

Construction: Construct chord $\overline{DE}$ such that $DE = AB$. Draw $\overline{OG}$ such that $\overline{OG} \perp \overline{DE}$. Draw segment $\overline{FG}$.

NOTE: The case we are showing here has O on the interior of inscribed angle $\angle CDE$. There are two other possibilities: O is on $\angle CDE$ and O is exterior to $\angle CDE$. These two cases can be verified separately. The first uses Theorem 6.36 and the second can be converted to the case shown below by drawing $\overline{DE}$ so O is inside $\angle CDE$.

Proof: STATEMENTS	REASONS
1. The circle centered at O with chords $\overline{AB}$ and $\overline{CD}$	1. Given
2. $DE = AB$	2. By construction
3. $\overline{OG} \perp \overline{DE}$	3. By construction
4. $\overline{OE} \perp \overline{AB}$ and $\overline{OF} \perp \overline{CD}$	4. Given
5. $OG = OE$	5. = chords are equidistant from center.
6. $OF > OE$	6. Given
7. $OF > OG$	7. Substitution law
8. $m\angle 3 > m\angle 4$	8. $\angle$'s opp. $\neq$ sides of $\triangle$ are $\neq$ in same order
9. $m\angle 3 + m\angle 2 = m\angle OGD$ and $m\angle 4 + m\angle 1 = m\angle OFD$	9. $\angle$ Add. Post.
10. $\angle OGD$ and $\angle OFD$ are right angles	10. $\perp$ lines form rt. $\angle$'s
11. $m\angle OGD = m\angle OFD$	11. All rt. $\angle$'s are = in measure
12. $m\angle 3 + m\angle 2 = m\angle 4 + m\angle 1$	12. Substitution law
13. $m\angle 2 < m\angle 1$	13. Subtraction property of inequalities with statements 8 and 12
14. $FD < DG$	14. Sides opp. $\neq \angle$'s are $\neq$ in same order

(Continued on the next page.)

Proof: STATEMENTS REASONS

15. $2FD < 2DG$ 15. Mult. prop. of ineq's

16. $CF = FD$ and $DG = GD$ 16. Line through $O \perp$ to chord bisects chord

17. $CD = CF + FD$ and $DE = DG + GE$ 17. Seg. Add. Post.

18. $CD = FD + FD$ and $DE = DG + DG$ 18. Substitution law

19. $CD = 2FD$ and $DE = 2DG$ 19. Distributive law

20. $CD < DE$ 20. Substitution law

21. $CD < AB$ 21. Substitution using statements 2 and 20

25. *Given:* $\triangle ABC$ inscribed in a circle centered at O with $m\angle A > m\angle B$.

Prove: $m\overset{\frown}{BC} > m\overset{\frown}{AC}$

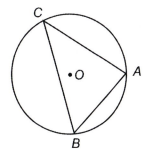

Proof: STATEMENTS REASONS

1. $\triangle ABC$ is inscribed in the circle centered at 0   1. Given

2. $m\angle A > m\angle B$ 2. Given

3. $BC > AC$ 3. Two $\angle$'s of $\triangle$ are $\neq$, sides opp. $\angle$'s are $\neq$ in same order.

4. $m\overset{\frown}{BC} > m\overset{\frown}{AC}$ 4. The $>$ of two chords has the $>$ arc (Theorem 6.32).

27. First prove that $RS = CD$ by drawing perpendiculars from O to $\overline{RS}$ and $\overline{CD}$ at points W and V respectively. Two right triangles are formed where $\overline{OP} \cong \overline{OP}$ by Reflexive law and $\angle 2 \cong \angle RPO$ by construction thus $\triangle OWP \cong \triangle OVP$ by HA. $m\angle 1 + m\angle RPA = m\angle RPO$ by Angle Addition Postulate. $m\angle 1 + m\angle RPA = m\angle 2$ by substitution thus $m\angle 1 > m\angle 2$.

Section 6.6 Locus and Basic Theorems (optional)

6.6 PRACTICE EXERCISES

1. The locus consists of all points on two concentric circles centered at O with radii $r - g$ and $r + g$.

2. Statement 1 is: $\overrightarrow{BD}$ is the bisector of $\angle ABC$
   The reason for Statement 2 is: Def. of $\angle$ bisector

   The reason for Statement 4 is: $\perp$ lines form rt. $\angle$'s

   The reason for Statement 5 is: Def. of rt. $\triangle$

   Statement 6 is: $\overline{BD} \cong \overline{BD}$

   Statement 8 is: $\overline{AD} \cong \overline{CD}$

   Statement 9 is: D is equidistant from $\overrightarrow{BA}$ and $\overrightarrow{BC}$.

6.6 SECTION EXERCISES

1. The locus is a circle with center P and radius 5 units.

3. The locus is two circles concentric with the given circle one with radius 1 unit and the other has radius 7 units.

5. The locus is a line between the given lines parallel to each of them at a distance 1.5 units from each.

7. There are three possibilities:
   (1) There is no locus if m does not intersect the circle.
   (2) The locus is one point, the point of tangency, if m is tangent to the circle.
   (3) The locus is two points if m intersects the circle in two points.

9. The locus is a diameter (excluding its endpoints) that is perpendicular to the given chords.

11. The locus of all points that are centers of circles tangent to both of two parallel lines is a line between the given parallel lines that is parallel to each. Thus, the desired locus is one point that is the intersection of the above line and the given intersecting line.

13. The proof consists of two parts. First we prove that the center of a circle tangent to both of two parallel lines is equidistant from the two given lines.

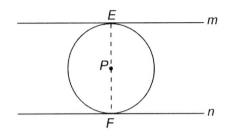

Given: P is the center of a circle tangent to m and n with m ∥ n

Prove: P is equidistant from m and n

Construction: Draw radii $\overline{PE}$ and $\overline{PF}$ to points of tangency E and F

Proof:

STATEMENTS	REASONS
1. P is the center of a circle tangent to m and n with m ∥ n	1. Given
2. $\overline{PE}$ and $\overline{PF}$ are radii to points of tangency E and F	2. By construction
3. $\overline{PE} \cong \overline{PF}$	3. Radii are ≅
4. $\overline{PE} \perp m$ and $\overline{PF} \perp n$	4. Radii to pts. of tangency are ⊥ to tangents
5. PE is distance from P to m and PF is distance from P to n	5. Def. of distance pt. to line
6. P is equidistant from m and n	6. By statements 3 and 5

Next we prove that a point on the line equidistant from two parallel lines is the center of a circle tangent to each line.

Given: P is on the line l that is equidistant from ∥ lines m and n.

Prove: P is the center of a circle tangent to both m and n.

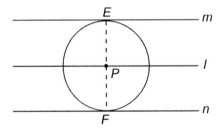

Construction: Draw $\overline{PE} \perp m$ and $\overline{PF} \perp n$

Proof:

STATEMENTS	REASONS
1. P is on l that is equidistant from parallel lines m and n.	1. Given
2. $\overline{PE} \perp m$ and $\overline{PF} \perp n$	2. By construction
3. PE = PF	3. Def. of equidistant
4. Construct circles centered at P with radius PE(= PF).	4. Circle can be constructed with given radius and center.
5. m and n are tangent to circle in step 4	5. Lines ⊥ to radii are tangent to circle

15. The locus consists of the point of intersection of the diagonals of the rectangle.

17. The locus of points is a line segment joining the center of the floor and ceiling. Think in three dimensions of a plane parallel to the front and back walls and half way between them. Now think of a second plane parallel to the side walls and half way between them. The intersection of the two planes is the desired locus of points.

Chapter 6 Review Exercises

1. The diameter is twice the radius or $2(2.6) = 5.2$ cm.

2. The radius is half the diameter of
$$\frac{1}{2}\left(\frac{4}{5}\right) = \frac{2}{5} \text{ yd.}$$

3. $m\angle P = 55°$ Since $\overline{PR}$ is a diameter, $\overarc{PQR}$ is a semicircle, $m\overarc{PQR} = 180°$.

$m\overarc{QR} = m\overarc{PQR} - m\overarc{PQ}$

thus $m\overarc{QR} = 180° - 70° = 110°$. $\angle P$ is an inscribed $\angle$ therefore it's measure is $\frac{1}{2}\overarc{QR}$.

$$m\angle P = \frac{1}{2}(110°) = 55°$$

4. $m < PQR = 90°$ because it is inscribed in a semicircle.

5. (a) Central angle
   (b) Inscribed angle

   (c) $\overarc{AB}$ has the same measure as central angle $\angle AOB$ which is $50°$.

   (d) $m\overarc{ABC} = 360° - m\overarc{AB} = 360° - 50° = 310°$
   (e) $m\angle ACB = \frac{1}{2}m\overarc{AB} = \frac{1}{2}(50°) = 25°$

6. (a) $90°$; because $\angle BCD$ is inscribed in a semicircle, it measures $90°$.

   (b) $55°$; because it is the intercepted arc of central $\angle AOB$, their measures are equal.

   (c) $27.5°$; because the measure of an inscribed angle is ½ the measure of the intercepted arc.

   (d) $60°$; because $\triangle OBC$ is equilateral hence equiangular. $m\angle BOC = 60°$, and $m\overarc{BC} = 60°$.

   (e) $120°$; because $\overline{BD}$ is a diameter, $m\overarc{BCD} = 180°$ and $m\overarc{DC} = m\overarc{BCD} - m\overarc{BC} = 180° - 60° = 120°$.

   (f) $120°$; because it is a central angle whose intercepted arc measures $120°$.

   (g) $30°$; The $\triangle ODC$ is isosceles (2 radii are congruent) therefore $m\angle ODC = m\angle OCD$. The sum of the measures of the angles of a triangle is $180°$. From part (f) $m\angle DOC = 120°$ thus $m\angle ODC = 30°$.

   (h) $125°$; because the sum of the arcs of a circle is $360°$. $360° - 120° - 60° - 55° = 125°$.

7. $m\angle BPC = \dfrac{1}{2}(m\overset{\frown}{BC} + m\overset{\frown}{AD})$

$\qquad = \dfrac{1}{2}(56° + 132°)$

$\qquad = \dfrac{1}{2}(188°) = 94°$

8. Since $CD = DA, m\overset{\frown}{CD} = m\overset{\frown}{DA}$. Since $m\overset{\frown}{CD} = 135°, m\overset{\frown}{DA} = 135°$

9. By Theorem 6.8, $CP = AP$ since the segment $\overline{BD}$ bisects chord $\overline{AC}$. Thus, $CP = 17$ cm.

10. By Theorem 6.13, $(CP)(AP) = (BP)(PD)$. Substituting we have $6)(5) = (3)(PD)$ making

$$PD = \frac{(6)(5)}{3} = 10 \text{ ft}$$

11. By Theorem 6.10, the distances to the center are equal.

12. $m\angle P = \dfrac{1}{2}(m\overset{\frown}{BD} - m\overset{\frown}{AC})$

$\qquad = \dfrac{1}{2}(68° - 28°)$

$\qquad = \dfrac{1}{2}(40°)$

$\qquad = 20°$

13. $m\angle P = \dfrac{1}{2}(m\overset{\frown}{BD} - m\overset{\frown}{AC})$ so substituting we have $24° = \dfrac{1}{2}(m\overset{\frown}{BD} - 30°)$ which gives $48° = m\overset{\frown}{BD} - 30°$ or $m\overset{\frown}{BD} = 78°$.

14. By Theorem 6.15, $(PB)(PA) = (PD)(PC)$. Substituting we have $(12)(3) = (PD)(4)$ so that

$$PD = \frac{(12)(3)}{4} = (3)(3) = 9 \text{ ft}$$

15. *Proof:*

STATEMENTS	REASONS
1. $ABCD$ is a rectangle	1. Given
2. $\overline{AB} \cong \overline{DC}$	2. Opp. sides of rect. are $\cong$
3. $\overset{\frown}{AB} \cong \overset{\frown}{CD}$	3. $\cong$ chords have $\cong$ arcs

16. *Proof:*

STATEMENTS	REASONS
1. $\overline{AD} \perp \overline{BC}$ and $\overline{AD}$ contains the center O	1. Given
2. $\overline{AD}$ bisects $\overline{BC}$	2. Line through center $\perp$ chord bisects the chord
3. $\overline{BD} \cong \overline{CD}$	3. Def. of bisector
4. $\angle ADB$ and $\angle ADC$ are right angles	4. $\perp$ lines form rt. $\angle$'s
5. $\triangle ADB$ and $\triangle ADC$ are right triangles	5. Def. of rt. $\triangle$
6. $\overline{AD} \cong \overline{AD}$	6. Reflexive law
7. $\triangle ADB \cong \triangle ADC$	7. LL
8. $\overline{AB} \cong \overline{AC}$	8. CPCTC
9. $\overset{\frown}{AB} \cong \overset{\frown}{AC}$	9. $\cong$ chords have $\cong$ arcs

17. $m\angle PAB = \frac{1}{2}m\widehat{AB} = \frac{1}{2}(66°) = 33°$ by Theorem 6.16.

18. Since $m\angle ADB = 35°$, $m\widehat{AB} = 2(35°) = 70°$ by Theorem 6.2. By Theorem 6.16,

$$m\angle PAB = \frac{1}{2}m\widehat{AB}$$
$$= \frac{1}{2}(70°)$$
$$= 35°$$

19. $m\angle APD = \frac{1}{2}(m\widehat{AD} - m\widehat{AB})$
$$= \frac{1}{2}(130° - 64°)$$
$$= \frac{1}{2}(66°)$$
$$= 33°$$

by Theorem 6.17.

20. Since $m\angle\widehat{ABC} = 130°$,

$m\widehat{ADC} = 360° - 130° = 230°$. By Theorem 6.18,

$$m\angle APC = \frac{1}{2}(m\widehat{APC} - m\widehat{ABC})$$
$$= \frac{1}{2}(230° - 130°)$$
$$= \frac{1}{2}(100°)$$
$$= 50°$$

21. By Theorem 6.19, $AP = CP$. Since $AP = 38$ cm, $CP = 38$ cm.

22. By Theorem 6.20, AP is the geometric mean between DP and BP. Thus,

$$\frac{DP}{AP} = \frac{AP}{BP}$$
$$\frac{30}{20} = \frac{20}{BP} \quad \text{Substitute 30 for } DP \text{ and 20 for } AP$$
$$(30)(BP) = (20)(2)$$
$$BP = \frac{(20)(20)}{30}$$
$$= 13.\overline{3} \text{ ft}$$

Remember that $13.\overline{3}$ corresponds to $13\frac{1}{3}$.

23. *Proof:*

STATEMENT	REASONS
1. $m\overset{\frown}{BD} = 2m\overset{\frown}{AC}$	1. Given
2. $m\angle B = \frac{1}{2}m\overset{\frown}{AC}$	2. Measure of inscribed $\angle$ is $\frac{1}{2}$ measure of intercepted arc
3. $m\angle P = \frac{1}{2}(m\overset{\frown}{BC} - m\overset{\frown}{AC})$	3. Theorem 6.14
4. $m\angle P = \frac{1}{2}(2m\overset{\frown}{AC} - m\overset{\frown}{AC})$	4. Substitution law
5. $m\angle P = \frac{1}{2}m\overset{\frown}{AC}$	5. Distributive law
6. $m\angle P = m\angle B$ thus $\angle P \cong \angle B$	6. Trans. law; Def. $\cong \angle$'s
7. $\overline{PC} \cong \overline{BC}$	7. Sides opp. $\cong \angle$'s are $\cong$

24. *Proof:*

STATEMENT	REASONS
1. $\overline{AD}$ and $\overline{DC}$ are tangents	1. Given
2. $AD = DC$	2. Theorem 6.19
3. $ABCD$ is a parallelogram	3. Given
4. $ABCD$ is a rhombus	4. Def. of rhombus

25. *Proof:*

STATEMENT	REASONS
1. $\overline{AB}$ is a common internal tangent	1. Given
2. $\overleftrightarrow{OP}$ is the line of centers	2. Given
3. $\overline{OA} \perp \overline{AB}$ and $\overline{PB} \perp \overline{AB}$	3. Radius to pt of tangency is $\perp$ to tangent
4. $\angle OAC$ and $\angle CBP$ are right angles	4. $\perp$ lines form rt. $\angle$'s
5. $\angle OAC \cong \angle PBC$	5. All rt. $\angle$'s are $\cong$
6. $\angle OCA \cong \angle PCB$	6. Vert. $\angle$'s are $\cong$
7. $\triangle OAC \sim \triangle PBC$	7. AA
8. $\angle O \cong \angle P$	8. Corr. $\angle$'s in $\sim\triangle$ are $\cong$

26. *Given:* Circles centered at O and P tangent externally at D, n is their common internal tangent, A is on n, $\overline{AC}$ is tangent to circle centered at O, and $\overline{AB}$ is tangent to circle centered at P.

Prove: $AC = AB$

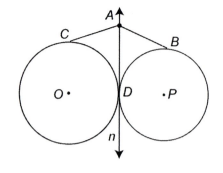

Proof: STATEMENTS   REASONS

1. n is common internal tangent to circles at D with A on n.   1. Given

2. $\overline{AC}$ is tangent to a circle centered at O and $\overline{AB}$ is tangent to circle centered at P.   2. Given

3. $AC = AD$ and $AD = AB$ 3. Tangents to circle from same pt. are =

4. $AC = AB$ 4. Trans. law

27. Use Construction 6.1

28. **(a)** $m\overset{\frown}{BC} = 100°$; $\dfrac{m\overset{\frown}{BC} - 20°}{2} = 40°$

(b) $m\overset{\frown}{AB} = 60°$;

$360° - (75° + 55° + 20° + 30° + 20°)$

(c) $m\angle EGD = 10°$; $\dfrac{1}{2}(75° - 55°)$

(d) $m\angle AGF = 5°$; $\dfrac{1}{2}(60° - 50°)$

(e) $m\angle BKC = 65°$; $\dfrac{1}{2}(100° + 30°)$

(f) $KC = 7.5$ cm; $1.4(KC) = 7(1.5)$

(g) $GD \approx 5.2$ cm;

$$\frac{2.7 + 7.4}{GD} = \frac{GD}{2.7}$$

$$\frac{10.1}{GD} = \frac{GD}{2.7}$$

$$(GD)^2 = 27.27$$

$$GD = \sqrt{27.27} \approx 5.2$$

(h) $AG \approx 5.2$ cm; By Theorem 6.19, 2 tangent segments from the same point, have the same length.

29. By Theorem 6.23, $\angle D$ and $\angle B$ are supplementary. Thus, $m\angle D + m\angle B = 180°$ so that $88° + m\angle B = 180°$ making $m\angle B = 180° - 88° = 92°$.

30. Since $\overline{AD} \parallel \overline{BC}$ and $AD = BC$, by Theorem 4.6, $ABCD$ is a parallelogram. By Theorem 6.24, $ABCD$ is a rectangle so that $m\angle D = 90°$.

31. Divide the circle into four equal arcs using the endpoints of a diameter and the points of intersection of the perpendicular bisector of the diameter and the circle. The perpendicular bisector of a side of the square determines the bisector of the arc formed by the sides which is used to mark off 8 equal arcs to form the regular octagon.

32. Construct the regular hexagon by inscribing it in a circle marking off arcs using the radius of the circle. Then use Construction 6.6 to inscribe a circle in the hexagon.

33. Each central angle of a regular 18-gon has measure $\frac{360°}{18} = 20°$.

34. Solve $10° = \frac{360°}{n}$ for n to obtain $n = 36$ sides.

35. The hexagon has sides equal in length to the radius of the circle which is 14 cm.

36. The following paragraph-style proof shows a regular hexagon but the same proof can be applied to any regular polygon with n sides where $n > 2$.

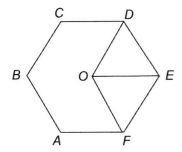

Proof:

Given: regular polygon *ABCDEF* where $\overline{OE}$ is a radius of the polygon

Prove: $\overline{OE}$ bisects $\angle DEF$

$\overline{OD} \cong \overline{OE} \cong \overline{OF}$ because all radii of a regular polygon are congruent. $\overline{DE} \cong \overline{EF}$ by the definition of a regular polygon, all sides are congruent. Thus $\triangle ODE \cong \triangle OFE$ by SSS.

$\angle DEO \cong \angle FEO$ by CPCTC thus $\overline{OE}$ bisects $\angle DEF$ by the definition of an angle bisector.

37. True; the greater of two central angles intercepts the greater of two arcs (Theorem 6.30).

38. False; since $m\angle 1 > m\angle 2$, $m\overparen{PR} > m\overparen{QS}$ by Theorem 6.3. By Theorem 6.33, $PR > QS$.

39. False; since $PQ > RS$, by Theorem 6.34, $\overline{PQ}$ is neared the center O than $\overline{RS}$ making $TO < UO$.

40. False; since $m\overparen{PQ} > m\overparen{RS}$, by Theorem 6.33, $PQ > RS$.

41. True; since $TO > UO$, $\overline{RS}$ is nearer center O than $\overline{PQ}$ so that $RS > PQ$ by Theorem 6.35. Then By Theorem 6.32, $m\overparen{RS} > m\overparen{PQ}$.

42. The locus is two circles concentric with the given circle and with radii 8 units and 12 units.

43. The locus is one circle concentric with the given circle with radius 10 units.

44. The locus is one point at the intersection of the bisector of $\angle ABC$ and the circle centered at *B*.

45. There are four possibilities: (1) There is no locus if *m* does not intersect the square. (2) The locus is one point if *m* passes through a vertex of the square only. (3) The locus is two points if *m* intersects two sides of the square. (4) The locus is a side of the square if *m* is collinear with that side.

Chapter 6 Practice Test

1. $r = 0.65$ ft; $d = 2r$ $1.3 = 2r$

2. $m\angle BAD = \frac{1}{2}(m\overset{\frown}{BD}) = \frac{1}{2}(130°) = 65°.$

3. $m\angle P = \frac{1}{2}(m\overset{\frown}{BD} - m\overset{\frown}{AC}).$ Since

$m\angle BOD = 130°,\ m\overset{\frown}{BD} = 130°.$ Since

$m\angle ADC = 30°,\ m\overset{\frown}{AC} = 2(30°) = 60°.$
Thus,

$$m\angle P = \frac{1}{2}(130° - 60°)$$
$$= \frac{1}{2}(70°) = 35°.$$

4. $(DE)(AE) = (BE)(EC)$ so $(DE)(4) = (6)(2)$

making $DE = \frac{(6)(2)}{4} = 3$ cm

5. $m\angle AEC = \frac{1}{2}(m\overset{\frown}{AC} + m\overset{\frown}{BD})$
$$= \frac{1}{2}(60° + 130°)$$
$$= \frac{1}{2}(190°)$$
$$= 95°.$$

6. Since $MW = NW$ and $MW = 4$ cm, $NW = 4$ cm.

7. $(PQ)(PR) = (PT)(PS)$ so substituting we have $(12)(6) = (9)(PS)$ which gives $PS = \frac{(12)(6)}{9} = 8$ cm.

8. Since $PA = PB$ and $PA = 15$ ft, $PB = 15$ ft.

9. Use Construction 6.2.

10. Since $\angle B$ and $\angle D$ are supplementary, and $m\angle B = 100°$, $m\angle D = 80°.$

11. Use Construction 2.4 and Construction 6.6.

12. The measure of each central angle of a regular octagon is $\frac{360°}{8} = 45°.$

13. $m\angle 1 > m\angle 2$ since the greater of two arcs is intercepted by the greater of two central angles.

14. $\overline{UV}$ is nearer center O since $m\overset{\frown}{XY} < m\overset{\frown}{UV}$ implies $XY < UV$ making $\overline{UV}$ nearer O.

15. $XY > UV$ since the greater of two central angles intercepts the greater of two arcs and the greater of two arcs has the greater of two chords.

16. *Given:* Rhombus *ABCD* with $m\angle A > 90°$ (not a square), and diagonals $\overline{AC}$ and $\overline{BD}$.

Prove: $AC \neq BD$

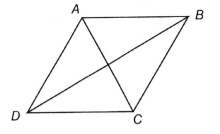

Proof: STATEMENTS   REASONS

STATEMENTS	REASONS
1. *ABCD* is a rhombus with $m\angle A > 90°$	1. Given
2. $m\angle A + m\angle D = 180°$	2. Adj. $\angle$'s in $\square$ are supp.
3. $m\angle D < 90°$	3. Subtraction property of inequalities
4. $m\angle A > m\angle D$	4. Transitive law
5. $\overline{AB} \cong \overline{AD} \cong \overline{DC}$	5. Sides of rhombus are $\cong$
6. $BD > AC$ (thus, $AC \neq BD$)	6. SAS Inequality Theorem

17. The locus is a line parallel to and between the given parallel lines and 3 units from each line.

18. There are three possibilities: (1) There is no locus if the circles do not intersect. (2) The locus is one point (the point of tangency) if the centers of the two circles are 2 units or 10 units apart. (3) The locus is two points if the circles intersect in two points.

19. The locus of the midpoints of all the chords parallel to a diameter is the perpendicular bisector of the diameter.

20. The locus is all points interior to a semicircle with radius 8 ft and centered at the midpoint of the fence.

CHAPTER 7   AREAS OF POLYGONS AND CIRCLES

Section 7.1 Areas of Quadrilaterals

7.1 PRACTICE EXERCISES

1. 1. $s = 20$ because $\sqrt{400} = 20$

2. 20 ft

3. 18 ft

4. $4(18) = 72$ ft

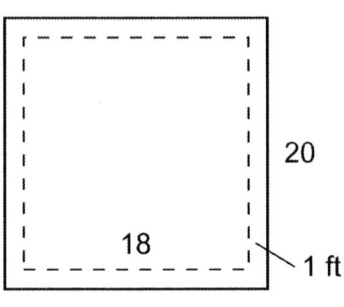

2. 1. $ABCD$ is a trapezoid

2. Addition property of areas

3. Formula for area $\triangle$

4. Area $\triangle BCD = \frac{1}{2}bh$

5. Substitution

6. Area $= \frac{1}{2}(b+b')h$

7.1 SECTION EXERCISES

1. $A = bh = (3)(4) = 12$ cm^2

$P = 3 + 3 + 4 + 4$

$\quad = 14$ cm

3. $A = bh = (10)(5) = 50$ yd^2

$P = 10 + 10 + 6 + 6 = 32$ yd

5. $A = s^2 = (11)^2 = 121$ cm^2

$P = 4s = 4(11) = 44$ cm

7. $A = \frac{1}{2}bh = \frac{1}{2}(12)(6) = (6)(6) = 36$ yd^2

$P = 7 + 11 + 12 = 30$ yd

9. $A = \frac{1}{2}bh = \frac{1}{2}(16)(4) = (8)(4) = 32$ cm^2

$P = 9 + 9 + 16 = 34$ cm (both legs are 9 cm)

11. $A = \frac{1}{2}(b+b')h = \frac{1}{2}(30+15)$

$\quad = \frac{1}{2}(45)(15) = 337.5$ yd^2

$P = 15 + 16 + 30 + 18 = 79$ yd

13. $A = \frac{1}{2}dd' = \frac{1}{2}(4.5)(6.7) = 15.075$ in^2

$P = 4.0 + 4.0 + 4.0 + 4.0 = 16.0$ in

15. $A = \frac{1}{2}(b+b')h = \frac{1}{2}(7.5+12)(3) = 29.25$ cm^2

$P = 7.5 + 3 + 2 + 5 = 27.5$ cm

17. Area of $\triangle = \frac{1}{2}bh = \frac{1}{2}(10)(4) = 20$ yd^2

Area of $\square = bh = (10)(6) = 60$ yd^2

Area of Figure = Area of $\triangle$ + Area of $\square$

$\quad = 20 + 60 = 80$ yd^2

$P = 7 + 7 + 6 + 10 + 6 = 36$ yd

19. The figure contains 2 right triangles and 1 equilateral $\triangle$. Use the Pythagorean Theorem to find the missing leg of the right $\triangle$.

$$3^2 + b^2 = 5^2; \; b^2 = 25 - 9 = 16$$
$$b = 4$$

The area of each right $\triangle = \dfrac{1}{2}(4)(3) = 6 \text{ ft}^2$.

$$\text{Area of equilateral } \triangle = \frac{5^2\sqrt{3}}{4} = \frac{25\sqrt{3}}{4}.$$

$$\text{Area of figure} = 6 + 6 + \frac{25\sqrt{3}}{4}$$
$$= \left(12 + \frac{25\sqrt{3}}{4}\right) \text{ ft}^2$$

21. 1. Given   2. $\overline{AC} \perp \overline{BD}$ 3. Add. prop. of area   4. Formula for area of $\triangle$
5. Area of $\triangle ABC = \frac{1}{2}(d)(BE)$
6. Substitution law   7. Distributive law
8. $DE + BE = BD = d'$ 9. $A = \frac{1}{2}dd'$

23. Use Heron's Formula

$$s = \frac{12 + 13 + 14}{2} = \frac{39}{2} = 19.5$$
$$A = \sqrt{19.5(19.5 - 12)(19.5 - 13)(19.5 - 14)}$$
$$= \sqrt{19.5(7.5)(6.5)(5.5)} = \sqrt{5228.4375}$$
$$\approx 72.3 \text{ in}^2$$

25. $A = bh = (18)(12) = 216 \text{ ft}^2$

27. In a right triangle the legs are perpendicular so one leg can be called the base and the other the altitude. Thus,

$$A = \frac{1}{2}bh + \frac{1}{2}(3.6)(5.2) = 9.36 \text{ ft}^2.$$

29. $A = \dfrac{1}{2}(b + b')h$

$$= \frac{1}{2}(6.3 + 11.5)(7.2)$$
$$= \frac{1}{2}(17.8)(7.2)$$
$$= 64.08 \text{ in}^2$$

31. Let h = height of rectangle. Since $A = bh$, we have $168 = 24h$. Divide both sides by 24 to get $h = 7$ cm.

33. Area of Wall $= (41)(9) - 2(2)(3)$
$$= 369 - 12 = 357 \text{ ft}^2$$

Since a gallon of paint covers 400 ft^2, to find the amount of paint needed for one coat divide 357 by 400. For two coats, multiply this quotient by 2. Thus,

$$\left(\frac{357}{400}\right)(2) = 1.785 \text{ gallons.}$$

Randy should buy 2 gallons of paint.

35. The patio is a rectangle 55 ft long and 20 ft wide. The area of the patio is

$$A = bh = (55)(20) = 1100 \text{ ft}^2.$$

Since 1 yd^2 = 9 ft^2, the area in yd^2 is

$$A = \frac{1100}{9} \text{ yd}^2.$$

Since each square yard of carpet costs $15.95, the project will cost

$$(15.95)\left(\frac{1100}{9}\right) \approx \$1949.44.$$

37. The driveway is a rectangle 32 ft wide and 24 ft long with area $A = lw = (24)(32) = 768$ ft^2. Since one gallon of sealer covers 300 ft^2, the number of gallons needed is

$$\frac{768}{300} \text{ gallons.}$$

Since one gallon of sealer costs \$15.95, the cost of the project is

$$\left(\frac{768}{300}\right)(15.95) = \$40.83.$$

39. Answers will vary.

41. The area of the shaded region is the area of the trapezoid minus the area of the rectangle. Thus, the area is

$$\frac{1}{2}(b+b')h - bh = \frac{1}{2}(6+11)(6) - (3)(4)$$

$$= \frac{1}{2}(17)(6) - 12$$

$$= 51 - 12 = 39 \text{ ft}^2$$

43. Let h = height of the rectangle. Then the length of the rectangle is $h + 2$. Since the area of a rectangle is $A = bh$, we must solve:

$$120 = (h+2)(h)$$

$$120 = h^2 + 2h \qquad \text{Distributive law}$$

$$0 = h^2 + 2h - 120 \qquad \text{Subtract 120}$$

$$0 = (h+12)(h-10) \quad \text{Factor}$$

$$h + 12 = 0 \text{ or } h - 10 = 0 \quad \text{Zero-product rule}$$

$$h = -12 \qquad h = 10$$

Since a height cannot be negative, we reject -12. Thus, the height is 10 ft and the base is 12 ft $(10 + 2 = 12)$.

45. By Corollary 5.14, CD is the mean proportional between AD and BD. Thus

$$\frac{AD}{CD} = \frac{CD}{BD}.$$

Substitute 4 for AD and 16 for BD.

$$\frac{4}{CD} = \frac{CD}{16}$$

$$(4)(16) = (CD)^2$$

$$64 = (CD)^2$$

$$\pm\sqrt{64} = CD$$

$$\pm 8 = CD$$

Reject the negative value -8, thus $CD = 8$. Since $AB = AD + BD = 4 + 16 = 20$ and $CD = 8$, the height of the triangle is 8 and the base is 20. Then

$$A = \frac{1}{2}bh = \frac{1}{2}(20)(8) = 10(8) = 80.$$

Thus, the area of the triangle is 80 cm^2.

47. In Exercise 45 we found $CD = 8$ cm. Since the area of $\triangle ADC$ is

$$A = \frac{1}{2}(AD)(CD)$$

we have

$$A = \frac{1}{2}(4)(8) = (2)(8) = 16 \text{ cm}^2.$$

Section 7.2 Circumference and Area of a Circle

7.2 PRACTICE EXERCISES

1. Since the circumference of a circle is given by
$$C = 2\pi r,$$
substitute 23.0 for C and solve for r.
$$23.0 = 2\pi r$$
$$\frac{23.0}{2\pi} = r$$
$$r \approx 7.3 \text{ ft}$$

2. Since $d = 2r$, substitute 6.4 for d and solve for r.
$$6.4 = 2r$$
$$3.2 = r$$
$$A = \pi r^2 = \pi(3.2)^2 = \pi(10.24)$$

Thus the area of the circle is 10.24π in^2.

3. First find the area of the triangle.
$$A_{\text{triangle}} = \frac{1}{2}bh = \frac{1}{2}(10)(5) = 25 \text{ in}^2$$

The area of each circular hole is
$$A_{\text{circle}} = \pi r^2 = \pi(1)^2 = \pi \text{ in}^2$$

The desired area is the area of the triangle minus three times the area of one circular hole.
$$A = 25 - 3\pi = 15.6 \text{ in}^2$$

7.2 SECTION EXERCISES

1. False; πr^2 is the formula for the area of a circle.

3. False; π is an irrational number; 3.14 is an approximation for π.

5. True

7. $C = 2\pi r = 2(6.2)\pi = 12.4\pi$ mi
$A = \pi r^2 = \pi(6.2)^2 = 38.44\pi$ mi^2

9. $C = \pi d = \pi(12.78) = 12.78\pi$ ft
$$A = \pi r^2 = \pi\left(\frac{12.78}{2}\right)^2$$
$$= \pi(6.39)^2 = 40.8321\pi \text{ ft}^2$$

11. $C = 2\pi r = 2\pi\left(\frac{3}{4}\right) = \pi\left(\frac{3}{2}\right) \approx 4.71$ cm;
$$A = \pi r^2 = \pi\left(\frac{3}{4}\right)^2 = \pi\left(\frac{9}{16}\right) \approx 1.77 \text{ cm}^2$$

13. Since $d = 2r$, $r = \frac{d}{2} = \frac{18.36}{2} = 9.18$ yd.
$$C = 2\pi r = \pi(9.18) \approx 57.68 \text{ yd};$$
$$A = \pi r^2 = \pi(9.18)^2 \approx 264.75 \text{ yd}^2$$

15. The area of the circle with a 10-meter radius is:
$$A = \pi r^2 = \pi(10)^2 = 100\pi \text{ m}^2$$
The area of the circle with a 9-meter radius is:
$$A = \pi r^2 = \pi(9)^2 = 81\pi \text{ m}^2$$

The area of the walk is the difference of these two areas.

Area of walk $= 100\pi - 81\pi = 19\pi \approx 59.69$ m^2

17. The area of the patio (including the dining area) is the area of the trapezoid with bases 8.1 yd and 6.7 yd, and height 5.8 yd.

$$A = \frac{1}{2}(b+b')h = \frac{1}{2}(8.1+6.7)(5.8)$$
$$= \frac{1}{2}(14.8)(5.8) = 42.92 \text{ yd}^2$$

The area of the dining area is the area of the circle with radius 1.6 yd (the diameter is 3.2 yd).

$$A = \pi r^2 = \pi(1.6)^2 = 256\pi \text{ yd}^2$$

The area to be carpeted is the difference of these two.

Area to carpet $= (42.92 - 2.56\pi)$ yd^2

The cost of the project is the cost per square yard ($18.50) times the number of square yards to be covered.

$$(42.92 - 2.56\pi)(18.50) = \$645.24$$

Remember, stores usually round up to the next penny.

19.   Area of Triangle $= \frac{1}{2}bh = \frac{1}{2}(9)(7.8)$
$$= 35.1 \text{ in}^2$$

Area of Each Circle $= \pi r^2 = \pi(1)^2 = \pi$ in^2

Shaded Area $=$ (Area of Triangle)
$$- 3(\text{Area of each Circle})$$
$$= 35.1 - 3\pi \approx 25.68 \text{ in}^2$$

21.   Area of Triangle $= \frac{1}{2}bh = \frac{1}{2}(22.5)(10.8)$
$$= 121.50 \text{ cm}^2$$

Area of Semicircle $= \frac{1}{2}\pi r^2 = \frac{1}{2}\pi(11.25)^2$
$$= 63.28125\pi \text{ cm}^2$$

Area of Rectangle $= bh = (44.5)(22.5)$
$$= 1001.25 \text{ cm}^2$$

Area of Each Circle $= \pi r^2 = \pi(2.6)^2$
$$= 6.76\pi$$

Area of Shaded Region $=$ (Area of Triangle)
$$+ (\text{Area of Semicircle})$$
$$+ (\text{Area of Rectangle})$$
$$- (\text{Area Each Circle})$$
$$= 121.50 + 63.28125\pi$$
$$+ 1001.25 - 3(6.76\pi)$$
$$\approx 1257.84 \text{ cm}^2$$

23. $C = \pi d$
$$\frac{5}{8}\pi = \pi d$$
$$\frac{5}{8} = d$$

Thus the diameter is $\frac{5}{8}$ in.

25. To find the area of the washer subtract the area of the inner circle from the outer circle.

$$\pi(2)^2 - \pi(0.5)^2 = 4\pi - 0.25\pi \approx 11.8 \text{ cm}^2$$

27. 52 inches is the tree's circumference $(C = 2\pi r)$.

$$\frac{52}{2\pi} = \frac{2\pi r}{2\pi} \text{ thus } r = \frac{52}{2\pi} \approx 8.3 \text{ inches}$$

We could say the tree's radius is approximately 8.3 inches.

29. The distance traveled by the earth in one day is $\frac{1}{365}$ times the circumference of the circular orbit. By Exercise 28, the distance traveled in the circular orbit in one year is

$2\pi(93,000,000)$. Thus, the distance traveled in one day is

$$\frac{1}{365}(2\pi)(93,000,000) \approx 1,600,000 \text{ miles.}$$

Section 7.3 Area and Arc Length of a Sector

1. By Postulate 7.6, the area of the colored sector is:

$$A = \frac{m}{360}\pi r^2$$

$$= \frac{90}{360}\pi(5)^2 \qquad \text{Substitute 90 for } m \text{ and 5 for } r$$

$$= \frac{1}{4}\pi(25) = \frac{25}{4}\pi \text{ cm}^2.$$

By Postulate 7.7, the length of the arc of the sector is:

$$L = \frac{m}{180}\pi r$$

$$= \frac{90}{180}\pi(5) \qquad \text{Substitute 90 for } m \text{ and 5 for } r$$

$$= \frac{1}{2}\pi(5) = \frac{5}{2}\pi \text{ cm.}$$

Perimeter of the region is $(\frac{5}{2}\pi + 10)$ cm.

3. By Postulate 7.6, the area of the colored sector is:

$$A = \frac{m}{360}\pi r^2$$

$$= \frac{45}{360}\pi(4)^2 \qquad \text{Substitute 45 for } m \text{ and 4 for } r$$

$$= \frac{1}{8}\pi(16) = 2\pi \text{ cm}^2.$$

By Postulate 7.7, the length of the arc of the sector is:

$$L = \frac{m}{180}\pi r$$

$$= \frac{45}{180}\pi(4) \qquad \text{Substitute 45 for } m \text{ and 4 for } r.$$

$$= \frac{1}{4}\pi(4) = \pi \text{ cm.}$$

Perimeter of the region is $(\pi + 8)$ cm.

5. The area of the colored segment is the area of the sector minus the area of the triangle. The area of the sector is:

$$A = \frac{m}{360}\pi r^2 = \frac{90}{360}\pi(4)^2 = \frac{1}{4}\pi(16)$$

$$= 4\pi \text{ yd}^2.$$

The area of the triangle is:

$$A = \frac{1}{2}bh = \frac{1}{2}(4)(4) = 8 \text{ yd}^2.$$

Thus, the area of the colored segment is:

$$4\pi - 8 = 4.6 \text{ yd}^2.$$

7. The area of the colored segment is the area of the sector minus the area of the triangle. The area of the sector is:

$$A = \frac{m}{360}\pi r^2 = \frac{60}{360}\pi(12)^2 = \frac{1}{6}\pi(144)$$

$$= 24\pi \text{ ft}^2.$$

The area of the triangle can be found by first drawing an altitude to the base which is also 12 ft, the same as the radius of the circle.

This forms a 30°-60°-90° triangle. The leg opposite the 60°-angle is the altitude and has length

$$\frac{\sqrt{3}}{2}(12) = 6\sqrt{3} \text{ ft by Theorem 5.20.}$$

Thus, the area of the triangle is:

$$A = \frac{1}{2}bh = \frac{1}{2}(12)(6\sqrt{3}) = 36\sqrt{3} \text{ ft.}$$

Thus, the area of the colored segment is:

$$24\pi - 36\sqrt{3} = 13.0 \text{ ft}^2.$$

9. The easiest method to use is to subtract the area of the 30°-60°-90° triangle from the area of the semicircle. Since the hypotenuse of the triangle is 16 cm (twice the radius of the circle). One leg is 8 cm and the other is

$$\frac{\sqrt{3}}{2}(16) = 8\sqrt{3} \text{ cm. The legs are the base and}$$

height of the triangle so its area is:

$$A = \frac{1}{2}bh = \frac{1}{2}(8)(8\sqrt{3}) = 32\sqrt{3} \text{ cm}^2.$$

Thus, the area of the colored region is:

$$32\pi - 32\sqrt{3} \approx 45.1 \text{ cm}^2.$$

11. Since the diameter is 8.2 cm, the radius is $\frac{8.2}{2} = 4.1$ cm. Substitute 4.1 for r and 30 for m in

$$A = \frac{m}{360}\pi r^2 = \frac{30}{360}\pi(4.1)^2$$

$$= \frac{1}{12}\pi(4.1)^2$$

$$\approx 4.4 \text{ cm}^2.$$

13. Substitute 24π for A and 60 for m into $A = \frac{m}{360}\pi r^2$ and solve for r.

$$24\pi = \frac{60}{360}\pi r^2$$

$$24 = \frac{1}{6}r^2 \quad \text{Divide both sides by } \pi.$$

$$144 = r^2 \quad \text{Multiply both sides by 6.}$$

$$\pm\sqrt{144} = r \quad \begin{array}{l}\text{Take square root}\\\text{of both sides.}\end{array}$$

$$\pm 12 = r$$

Reject the negative value, -12. Thus, the radius is 12 yd making the diameter twice this value or 24 yd.

15. $A = \dfrac{m}{360}\pi r^2$

$$A = \frac{25}{360}\pi(12.6)^2$$

$$A \approx 34.6 \text{ in}^2$$

17. $A = \dfrac{m}{360}\pi r^2$

$$A = \frac{12}{360}(720)$$

$$A = 24 \text{ cm}^2$$

19. The arc length of the segment is needed to find the perimeter of the segment.

$$L = \frac{m}{360}2\pi r$$

$$L = \frac{90}{360}2\pi(10)$$

$$L = 5\pi$$

The perimeter of the segment is equal to the sum of the arc length and the two radii.

$$P = (5\pi + 20) \text{ cm}$$

To find the area of the segment, subtract the area of the triangle from the area of the sector. The area of the sector is

$$= \frac{m}{360}\pi r$$

$$= \frac{90}{360}\pi(10)^2$$

$$= \frac{1}{4}\pi(100)$$

$$= 25\pi.$$

Since the triangle is a right triangle, the legs of the triangle are the base and height.

$$A = \frac{1}{2}bh$$

$$= \frac{1}{2}(10)(10)$$

$$= 50$$

The area of the segment is $(25\pi - 50)$ cm^2.

21. The shaded region can be thought of as two segments (formed by a diagonal of the square). The area of each segment is the area of the sector with radius 4 yd and arc 90° minus the area of the triangle with base 4 yd and height 4 yd. Thus, we have

$$\frac{m}{360}\pi r^2 - \frac{1}{2}bh = \frac{90}{360}\pi(4)^2 - \frac{1}{2}(4)(4)$$

$$= \frac{1}{4}\pi(16) - 8$$

$$= 4\pi - 8.$$

Since the shaded region is twice this area, the area of the shaded region is

$$2(4\pi - 8) = 9.1 \text{ yd}^2.$$

23. The distance around the top of the cone is the circumference.

$$C = 2\pi r$$

$$9.4 = 2\pi r$$

$$\frac{9.4}{2\pi} = r$$

$$r \approx 1.5 \text{ in.}$$

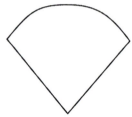

The circumference around the top of the cone corresponds to the arc length of the sector formed by the flattened cone.

Section 7.4 Area of Regular Polygons

7.4 PRACTICE EXERCISE

1. Use $A = \frac{1}{2}ap$ to find the area of the regular pentagon. $p = 5(6.5) = 32.5$ inches because a regular pentagon has 5 congruent sides.

$$A = \frac{1}{2}(4.5)(32.5)$$

$$A = 73.125 \text{ in}^2$$

7.4 SECTION EXERCISES

1. $A = \dfrac{1}{2}ap$

$A = \dfrac{1}{2}(10)(80)$

$A = 400 \text{ yd}^2$

3. A drawing may help to solve the problem.

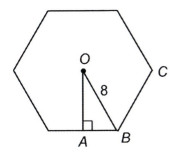

In the drawing $\overline{OA}$ is an apothem and $\overline{OB}$ is a radius of the regular hexagon. $\overline{OA} \perp \overline{AB}$ by the definition of apothem $OB = 8$ in. is given.

By Theorem 3.16 $m\angle ABC = \dfrac{(6-2)180°}{6}$

$= 120°$

By Theorem 7.10 $m\angle ABO = \dfrac{120°}{2}$

$= 60°$

Thus $\triangle AOB$ is 30°-60°-90° triangle where the hypotenuse = 8 in.

By Theorem 5.20, $AB = 4$, $OA = 4\sqrt{3}$

The apothem is $4\sqrt{3}$ inches and a side of the hexagon is $2(4) = 8$ inches.

5. A drawing may help solve the problem.

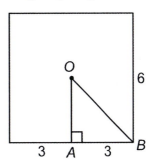

In the drawing $\overline{OA}$ is an apothem and $\overline{OB}$ is a radius. $\overline{OA}$ bisects the side of the square by Theorem 7.9 and $m\angle OBA = 45°$ by Theorem 7.10. $\triangle AOB$ is a 45°-45°-90° $\triangle$ thus $OA = 3$ and $OB = 3\sqrt{2}$. The apothem = 3 cm and the radius $= 3\sqrt{2}$ cm.

7. A drawing may help to solve the problem.

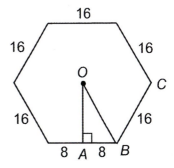

The length of the apothem (OA) is needed.

$\triangle AOB$ is 30°-60°-90° $\triangle$ (See problem 3 for a more detailed explanation.)

$AB = 8$ by Theorem 7.9 thus $OA = 8\sqrt{3}$ by Theorem 5.20.

$$A = \dfrac{1}{2}ap$$

$$A = \dfrac{1}{2}\left(8\sqrt{3}\right)(16 \cdot 6)$$

$$A = 384\sqrt{3} \text{ ft}^2$$

9. A drawing may help to solve the problem.

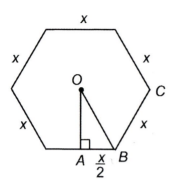

Let x represent a side of the regular hexagon.

$$AB = \frac{x}{2} \text{ by Theorem 7.9}$$

$\triangle AOB$ is 30°-60°-90°$\triangle$ (See problem 3 for a more detailed explanation.)

$$OA = \frac{x}{2}\sqrt{3} \text{ by Theorem 5.20}$$

$$A = \frac{1}{2}ap$$

$$1350\sqrt{3} = \frac{1}{2}\left(\frac{x}{2}\sqrt{3}\right)(6x)$$

$$1350\sqrt{3} = \frac{6\sqrt{3}x^2}{4}$$

$$1350\sqrt{3} = \frac{3\sqrt{3}x^2}{2}$$

$$\frac{2}{3\sqrt{3}}\left(1350\sqrt{3}\right) = x^2$$

$$900 = x^2$$

$$\pm 30 = x$$

Reject the negative value, −30. Thus each side of the regular hexagon is 30 cm.

11. A drawing may help to solve the problem.

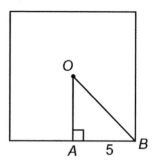

By Theorem 7.10, $m\angle ABO = 45°$ thus $\triangle AOB$ is a 45°-45°-90° triangle and $AB = 5$. One side of the square is 10 cm.

$$A = \frac{1}{2}ap$$

$$= \frac{1}{2}(5)(40)$$

$$= 100 \text{ cm}^2$$

13. Use the formula: $A = \frac{1}{2}ap$.

$$A = \frac{1}{2}(5.76)(6\cdot 8)$$

$$= \frac{1}{2}(5.76)(48)$$

$$= 138.24 \text{ cm}^2$$

15. The area of a regular 16-gon with apothem 12 ft can be approximated by the area of the inscribed circle with radius 12 ft.

$$A = \pi r^2 = \pi (12)^2 \approx 452.4 \text{ ft}^2$$

17. Use the formula: $A = \frac{1}{2}ap$.

$$338.6 = \frac{1}{2}(9)p$$

$$777.2 = 9p$$

$$p \approx 86.4 \text{ m}$$

19. Use the formula: $A = \frac{1}{2}ap$.

$$85 = \frac{1}{2}a(8)(4.25)$$

$$85 = \frac{1}{2}a(34)$$

$$170 = 34a$$

$$a = 5 \text{ in}$$

21. Use the formula: $A = \frac{1}{2}ap$ to find the area of

the entire hexagon.

$$A = \frac{1}{2}(8.7)(6 \cdot 10)$$

$$= \frac{1}{2}(8.7)(60)$$

$$= 261 \text{ mm}^2$$

The shaded area is $\frac{1}{3}$ of the total hexagon.

$$\frac{1}{3}(261) = 87 \text{ mm}^2$$

23. Answers may vary.

Chapter 7 Review Exercises

1. Perimeter is the sum of the sides of the figure.

$$P = 12 + 15 + 19 = 46 \text{ cm}$$

Use Heron's formula to find the area of the triangle since the height is not known.

$$A = \sqrt{s(s-a)(s-b)(s-c)}$$

where $s = \dfrac{a+b+c}{2}$

$$s = \frac{12+15+19}{2} = \frac{46}{2} = 23$$

$$A = \sqrt{23(23-12)(23-15)(23-19)}$$

$$= \sqrt{23(11)(8)(4)}$$

$$= \sqrt{8096} \approx 89.98 \text{ cm}^2$$

2. Perimeter is the sum of the sides of the figure.

$$P = 3 + 12.5 + 5 + 16.5 = 37 \text{ in.}$$

The formula for the area of a trapezoid is

$$A = \frac{1}{2}(b+b')h.$$

$$A = \frac{1}{2}(12.5+16.5)(3)$$

$$= \frac{1}{2}(29)(3)$$

$$= 43.5 \text{ in}^2$$

3. The sides of an equilateral triangle are equal in measure. $3s = P$ where s is a side of the triangle.

$$3s = 36$$
$$s = 12$$

The formula for the area of an equilateral triangle is $A = \dfrac{a^2 \sqrt{3}}{4}$.

$$A = \dfrac{(12)^2 \sqrt{3}}{4}$$
$$= \dfrac{144\sqrt{3}}{4}$$
$$= 36\sqrt{3} \text{ in}^2$$

4. The area of a square is $A = s^2$. If s is doubled, $s = 2s$ and the area is $A = (2s)^2 = 4s^2$ which is 4 times the original area.

5. The area of each wall is $A = (14)(8) = 112 \text{ ft}^2$. Two of the walls have windows measuring 4 ft by 3 ft so the area of each of these is

$$112 - (3)(4) = 112 - 12 = 100 \text{ ft}^2$$

The total area to be painted is

$$112 + 112 + 100 + 100 = 424 \text{ ft}^2.$$

Since 1 gallon of paint covers 300 ft^2, for one coat it would take

$$\dfrac{424}{300} \text{ gallons of paint.}$$

For two coats it would take

$$2\left(\dfrac{424}{300}\right) \approx 3 \text{ gallons of paint.}$$

6. The area of a rhombus with diagonals d and d' is given by:

$$A = \dfrac{1}{2}dd' = \dfrac{1}{2}(14)(18) = 126 \text{ cm}^2$$

7. Area of the left triangle is

$$\dfrac{1}{2}(30)(30) = 450 \text{ cm}^2.$$

Area of the trapezoid is

$$\dfrac{1}{2}(40 + 70)(27) = 1485 \text{ cm}^2.$$

The base angles of the trapezoid are congruent thus it is an isosceles trapezoid. Because of this fact, the length of the rectangle is 30 cm.

Area of the rectangle is $(20)(30) = 600 \text{ cm}^2$.

Area of the entire figure

$$= 450 + 1485 + 600$$
$$= 2535 \text{ cm}^2.$$

$$\text{Perimeter} = 40 + 43 + 30 + 70 + 20 + 30 + 20$$
$$= 253 \text{ cm.}$$

8. In a circle, $d = 2r$ thus $d = 2(2.6) = 5.2$ cm.

9. In a circle, $d = 2r$ thus $\dfrac{4}{5} = 2r$ and $r = \dfrac{2}{5}$ yd.

10. The formula for the area of a circle is $A = \pi r^2$ where $d = 2r$ thus $\dfrac{5}{8} = 2r$ and $r = \dfrac{5}{16}$ m.

$$A = \pi\left(\dfrac{5}{16}\right)^2$$
$$= \dfrac{25\pi}{256} \text{ m}^2$$

The formula for the circumference of a circle is $C = \pi d$.

$$C = \pi\left(\dfrac{5}{8}\right) \text{ or } \dfrac{5\pi}{8} \text{ m.}$$

11. Since the diameter is 9.2 ft, the radius is

$$\frac{9.2}{2} = 4.6 \text{ ft.}$$

The formula for the area of a circle is $A = \pi r^2$.

$$A = \pi(4.6)^2 \approx 66.48 \text{ ft}^2$$
$$C = 2\pi r = 2\pi(4.6) \approx 28.90 \text{ ft}$$

or $C = \pi d = \pi(9.2) \approx 28.90 \text{ ft}$

12. To find the area of the remaining metal, subtract the area of the circle from the area of the triangle.

Area of the equilateral triangle is

$$\frac{a^2\sqrt{3}}{4} = \frac{(10)^2\sqrt{3}}{4} = \frac{100\sqrt{3}}{4}.$$

Area of the circle is

$$\pi r^2 = \pi(1.5)^2 = 2.25\pi.$$

Area of the remaining metal is

$$\frac{100\sqrt{3}}{4} - 2.25\pi \approx 36.2 \text{ in}^2$$

13. To find the area of the CD, subtract the area of the CD from the area of the hole.

Since the diameter of the hole is $\frac{3}{2}$ cm, the radius is $\frac{3}{4}$ or 0.75 cm.

The area of the CD is

$$\pi(6)^2 = 36\pi \text{ cm}^2.$$

The area of the hole is

$$\pi(0.75)^2 = 0.5625\pi \text{ cm}^2.$$

The area of the surface of the CD is

$$36\pi - 0.5625\pi \approx 111.3 \text{ cm}^2.$$

14. $A = \frac{m}{360}\pi r^2 = \frac{18}{360}\pi(11.4)^2 \approx 20.4 \text{ in}^2$

15. The area of the segment is the area of the sector minus the area of the equilateral triangle with base 10 cm and altitude of $\frac{\sqrt{3}}{2}(10)$ cm.

Area of the triangle

$$= \frac{1}{2}(10)\left(\frac{\sqrt{3}}{2}(10)\right) = 25\sqrt{3}$$

Area of the sector

$$= \frac{60}{360}\pi(10)^2 = \frac{50}{3}\pi$$

Area of the segment

$$= \frac{50}{3}\pi - 25\sqrt{3} \approx 9.1 \text{ cm}^2$$

16. The formula for the length of an arc is $L = \frac{m}{360}2\pi r.$

$$L = \frac{40}{360}2\pi(5.2) = \frac{1}{9}\pi(10.4) \approx 3.6 \text{ ft}$$

17. To find the area of the shaded sector use the formula $A = \frac{m}{360}\pi r^2.$

$$A = \frac{60}{360}\pi(6)^2 = \frac{1}{6}\pi(36) = 6\pi \text{ cm}^2$$

To find the perimeter of the shaded sector, find the arc length and add the two radii.

$$L = \frac{m}{360}2\pi r = \frac{60}{360}2\pi(6)$$
$$= \frac{1}{6}(12)\pi = 2\pi$$

Perimeter $= 2\pi + 2(6) = (2\pi + 12) \text{ cm}$

18. To find the area cleaned, subtract the area of the smaller sector ($r = 8$ in) from the area of the entire sector ($r = 20$ in).

Area of smaller sector $= \dfrac{120}{360}\pi(8)^2$

$= \dfrac{1}{3}\pi(64)$

Area of entire sector $= \dfrac{120}{360}\pi(20)^2$

$= \dfrac{1}{3}\pi(400)$

Area cleaned by the wiper blade

$= \dfrac{1}{3}\pi(400) - \dfrac{1}{3}\pi(64)$

$= \dfrac{1}{3}\pi(336) = 112\pi \text{ in}^2$

19. The formula for the area of a regular polygon is $A = \dfrac{1}{2}ap$.

The perimeter of the square is $4(12) = 48$ cm.

$A = \dfrac{1}{2}(6)(48) = 144 \text{ cm}^2$

20. The formula for the area of a regular polygon is $A = \dfrac{1}{2}ap$.

$A = \dfrac{1}{2}(7)(120) = 420 \text{ ft}^2$

21. The formula for the area of a regular polygon is $A = \dfrac{1}{2}ap$.

The perimeter of the sign is $8(12.5) = 100$ in.

To find the apothem consider a right triangle formed by a radius and an apothem of the octagon.

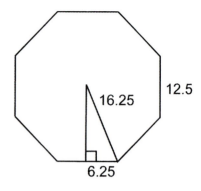

Using the Pythagorean theorem

$16.25^2 = x^2 + 6.25^2$

where x is the length of the apothem.

$x^2 = 225$ thus $x = 15$.

$A = \dfrac{1}{2}(15)(100) = 750 \text{ in}^2$

22. The formula for the area of a regular polygon is $A = \dfrac{1}{2}ap$.

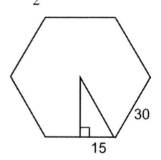

The perimeter of the hexagon is $6(30) = 180$ yd.

To find the apothem of the hexagon consider the 30°-60°-90° triangle formed by an apothem and a radius. The short leg is 15 thus the long leg (the apothem) is $15\sqrt{3}$.

$A = \dfrac{1}{2}\left(15\sqrt{3}\right)(180) \approx 2338.3 \text{ yd}^2$

23. Approximate the area of this figure by the area of a circle with radius 15.5 ft.

$A = \pi(15.5)^2 \approx 754.8 \text{ ft}^2$

Chapter 7 Practice Test

1. The formula for the area of a parallelogram is $A = bh$.

$$A = (20)(17) = 340 \text{ ft}^2$$

2. The formula for the area of a triangle is $A = \dfrac{1}{2}bh$.

$$60 = \frac{1}{2}(15)h$$
$$120 = 15h$$
$$8 = h$$

Thus the height is 8 yd.

3. The missing measurement on the width of the room is 3.0 yd since 3.0 + 3.0 = 6.0 (the opposite side of the rectangular room). The area of the room is the sum of the areas of the two rectangles that are 8.8 by 3.0 and 5.3 by 3.0.

$$(8.8)(3.0) + (5.3)(3.0) = 26.4 + 15.9$$
$$= 42.3 \text{ yd}^2$$

Subtract the area of the fire pit from the area of the room.

$$42.3 - (1.5)^2 = 40.05 \text{ yd}^2$$

If the carpeting costs $31.95 a square yard installed, the project cost is

$$40.05(31.95) = \$1279.60.$$

4. Use Heron's formula to find the area of the triangle.

$$A = \sqrt{s(s-a)(s-b)(s-c)} \text{ where } s = \frac{a+b+c}{2}.$$

$$s = \frac{5+6+7}{2} = 9;$$
$$A = \sqrt{9(9-5)(9-6)(9-7)}$$
$$= \sqrt{216}$$
$$\approx 14.7 \text{ mm}^2$$

5. The formula for the area of a rhombus is $A = \dfrac{1}{2}dd'$.

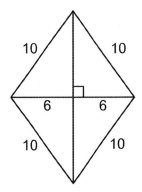

Since only one diagonal is given, the other one will be found using the Pythagorean Theorem. All sides of a rhombus are congruent. The perimeter is 40 inches thus each side measures 10 inches. The one diagonal is the perpendicular bisector of the other.

$$6^2 + x^2 = 10^2$$
$$36 + x^2 = 100$$
$$x^2 = 64$$
$$x = 8$$

8 inches is one half the other diagonal, thus

$$A = \frac{1}{2}(12)(16) = 96 \text{ in}^2.$$

6. The formula for the area of a trapezoid is

$$A = \frac{1}{2}(b + b')h.$$

$$40 = \frac{1}{2}(13 + 7)h$$

$$80 = 20h$$

$$4 = h$$

Thus the height measures 4 inches.

7. The formula for the area of a parallelogram is $A = bh$.

$$A = (18)(2) = 36 \text{ in}^2$$

The formula for the area of a triangle is

$$A = \frac{1}{2}bh.$$

$$A = \frac{1}{2}(12)(6) = 36 \text{ in}^2$$

The area of the two figures are equal but the figures are not congruent. Congruent figures are figures that can be made to coincide. These figures are not the exact same shape therefore will not coincide even though the areas are equal.

8. The formula for the area of an equilateral triangle is

$$A = \frac{a^2\sqrt{3}}{4}.$$

$$A = \frac{(33)^2\sqrt{3}}{4} = \frac{1089\sqrt{3}}{4}$$

$$\approx 471.6 \text{ in}^2$$

9. The formula for circumference of a circle is $C = \pi d$ or $2\pi d$.

$$C = \pi(12.6) \approx 39.58 \text{ cm}$$

The formula for the area of a circle is $A = \pi r^2$.

$$A = \pi\left(\frac{12.6}{2}\right)^2 = \pi(39.69)$$

$$\approx 124.69 \text{ cm}^2$$

10. To find the area of the shaded region, subtract the sum of the areas of the 4 sectors from the area of the square. The area of the four sectors area equal and $m = 90°$ while $r = 6$. The formula for the area of a sector is

$$A = \frac{m}{360}\pi r^2.$$

$$A = \frac{90}{360}\pi(6)^2 = 9\pi$$

thus the area of the 4 sectors is $4(9\pi) = 36\pi$.

The formula for the area of a square is $A = s^2$. The area of this square is $(12)^2 = 144$. The area of the shaded region is

$$144 - 36\pi \approx 30.9 \text{ in}^2.$$

11. To find the circumference of a circle the radius is needed. The formula for the area of a circle is $A = \pi r^2$.

$$25\pi = \pi r^2$$

$$25 = r^2$$

$$\pm 5 = r \quad (\text{reject } r = -5)$$

Thus $C = 2\pi(5) = 10\pi$ cm.

12. To find the area of the shaded region, subtract the area of the triangle from the area of the circle. It is a right triangle because it is inscribed in a semicircle. Use the Pythagorean Theorem to find the diameter and then the radius of the circle. $(12)^2 + (16)^2 = d^2$ where d is the diameter. $d^2 = 400$ thus $d = 20$ and $r = 10$ cm.

The area of the circle is

$$A = \pi(10)^2 = \pi(100).$$

The area of the triangle is

$$A = \frac{1}{2}(12)(16) = 96.$$

The area of the shaded region is

$$100\pi - 96 \approx 218.2 \text{ cm}^2.$$

13. The number of degrees for the angle in the shaded region is $360° - 225° = 135°$. The formula for the area of a sector is

$$A = \frac{m}{360}\pi r^2.$$

$$A = \frac{135}{360}\pi(12)^2 = \frac{3}{8}\pi(144) = 54\pi \text{ m}^2$$

Arc length is needed to find the perimeter of the sector.

The formula for arc length is $L = \frac{m}{360}2\pi r.$

$$L = \frac{135}{360}2\pi(12) = \frac{3}{8}\pi(24) = 9\pi$$

Perimeter is $9\pi + 12 + 12 = (9\pi + 24)$ m.

14. To find the arc length of the fan, use the formula

$$L = \frac{m}{360}2\pi r.$$

$$L = \frac{120}{360}2\pi(18) = 12\pi \text{ cm}$$

15. The formula for the area of a regular polygon is

$$A = \frac{1}{2}ap.$$

$$A = \frac{1}{2}(7)(56) = 196 \text{ yd}^2$$

16. To find the area of the segment, subtract the area of the triangle from the area of the sector. A drawing may help solve the problem.

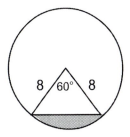

Since the central angle is 60°, the triangle is equilateral.

The area of the triangle is $\dfrac{(8)^2\sqrt{3}}{4} = 16\sqrt{3}.$

The area of the sector is $\dfrac{60}{360}\pi(8)^2 = \dfrac{32\pi}{3}.$

The area of the segment is

$$\frac{32\pi}{3} - 16\sqrt{3} \approx 5.8 \text{ cm}^2.$$

17. The formula for the area of a regular polygon is $A = \dfrac{1}{2}ap.$

The perimeter of the hexagon is $6(16) = 96$ ft.

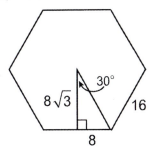

To find the apothem of the hexagon, consider the 30°-60°-90° triangle formed by an apothem and a radius. The short leg is 8 thus the long leg is $8\sqrt{3}$.

$$A = \frac{1}{2}\left(8\sqrt{3}\right)(96) = 384\sqrt{3} \text{ ft}^2$$

18. Approximate the area of this figure by the area of the inscribed circle with radius 12.2 ft.

$$A = \pi\left(12.2\right)^2 \approx 467.6 \text{ cm}^2$$

CHAPTERS 4-7 CUMULATIVE REVIEW

1. False   **2.** True   **3.** False   **4.** True   **5.** True   **6.** True

7. Consecutive angles of a parallelogram are supplementary.

$$x + 10 + 2x - 4 = 180$$
$$3x + 6 = 180$$
$$3x = 174$$
$$x = 58$$

Thus $m\angle A = 58 + 10 = 68° = m\angle C$

$m\angle B = 2(58) - 4 = 112° = m\angle D$

8. By Theorem 4.22, the median of a trapezoid is equal to one-half the sum of the bases.

$$\frac{3x + 1 + 12x - 2}{2} = 6x + 7$$
$$\frac{15x - 1}{2} = 6x + 7$$
$$15x - 1 = 12x + 14$$
$$3x = 15$$
$$x = 5$$

9. *Given:* $\triangle ABC$ and $\overline{AB} \parallel \overline{DE}$

Prove: $\dfrac{EC}{BC} = \dfrac{DC}{AC}$

Proof:

STATEMENTS	REASONS
1. $\triangle ABC$ and $\overline{AB} \parallel \overline{DE}$	1. Given
2. $\angle ABC \cong \angle DEC$	2. If lines $\parallel$, corresp. $\angle$'s $\cong$
3. $\angle BAC \cong \angle EDC$	3. If lines $\parallel$, corresp. $\angle$'s $\cong$
4. $\triangle ABC \sim \triangle DEC$	4. AA
5. $\dfrac{EC}{BD} = \dfrac{DC}{AC}$	5. Def ~ $\triangle$'s

10. By Theorem 5.16, $AE = 16$ cm

11. Use Corollary 5.14

$$\frac{AD}{CD} = \frac{CD}{DB}$$
$$\frac{16}{x} = \frac{x}{4}$$
$$x^2 = 64$$
$$x = 8 \quad \text{(reject the negative value)}$$

Thus $CD = 8$ in.

12. Use the Pythagorean Theorem and let $x = CB$.

$$16^2 + x^2 = 34^2$$
$$256 + x^2 = 1156$$
$$x^2 = 900$$
$$x = 30 \text{ in (reject the negative value)}$$

13. Use Theorem 5.20.
The short leg is 10 mm thus the hypotenuse is 20 mm and the long leg is $10\sqrt{3}$ mm.

14. Use Theorem 5.19 to find the length of one leg = 8 m.

$$A = \frac{1}{2}bh = \frac{1}{2}(8)(8) = 32 \text{ m}^2$$

15. False; If the measure of two angles of a triangle are unequal, the measures of the sides opposite those angles are unequal in the same order (Theorem 5.23). Since $m\angle A > m\angle B$ then $BC > AC$.

16. 40°   **17.** 20°   **18.** 140°

19. 90°   **20.** 30°   **21.** 60°

22. 120°   **23.** 70°   **24.** 120°

25. By Theorem 6.33 the greater of two arcs has the greater chord.

26. By Corollary 6.36, a diameter is greater than any other chord.

27. Use Theorem 6.20. Let $x = GD$.

$$\frac{AG}{GD} = \frac{GD}{CG}$$
$$\frac{20}{x} = \frac{x}{5}$$
$$x^2 = 100$$
$$x = 10$$

Thus $GD = 10$.

28. Use Theorem 6.29. The measure of each central angle is $\frac{360°}{8} = 45°$.

29. True by Theorem 6.32.

30. There are no points in the locus if line m does not intersect the pentagon. There is one point in the locus if line m only goes through one vertex of the pentagon. There are two points in the locus if line m intersects 2 sides of the pentagon. If line m coincides with one side of the pentagon, the locus consists of all the points on that side of the pentagon.

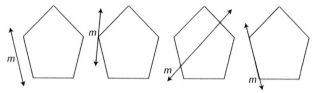

31. The formula for the area of a triangle is $A = \frac{1}{2}bh$.

$$60 = \frac{1}{2}(15)h$$
$$120 = 15h$$
$$8 = h$$

Thus the height of the triangle is 8 yd.

32. The formula for the circumference of a circle is $C = 2\pi r$.

$$C = 2\pi(25) = 50\pi \text{ in}^2$$

33. The area of the segment is the area of the sector minus the area of the equilateral triangle with base 20 cm and altitude of
$$\frac{\sqrt{3}}{2}(20) = 10\sqrt{3} \text{ cm.}$$

$$\text{Area of the triangle} = \frac{1}{2}(20)(10\sqrt{3})$$
$$= 100\sqrt{3}$$

$$\text{Area of the sector} = \frac{60}{360}\pi(20)^2$$
$$= \frac{200}{3}\pi$$

$$\text{Area of the segment} = \frac{200}{3}\pi - 100\sqrt{3}$$

$$\approx 36.2 \text{ mm}^2$$

34. Use the formula $A = \dfrac{1}{2}ap$.

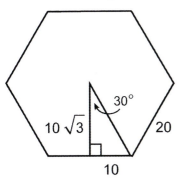

The perimeter of the hexagon is 6(20) = 120. To find the apothem, consider the 30°-60°-90° right triangle formed by the radius and apothem of the hexagon. The short leg is 10 thus the long leg (apothem) is $10\sqrt{3}$.

$$A = \frac{1}{2}\left(10\sqrt{3}\right)(120) = 600\sqrt{3} \text{ yd}^2$$

35. The formula for the area of a sector is

$$A = \frac{m}{360}\pi r^2.$$

$$A = \frac{36}{360}\pi(12.8)^2 \approx 51.5 \text{ in}^2$$

36. The formula for the area of a square is $A = s^2$. Since $s^2 = 36$, $s = \pm6$. Reject $s = -6$, thus one side of the square is 6 inches. The perimeter of the square is 4(6) = 24 inches. Thus the perimeter of the equilateral triangle is 24 inches so one side of the triangle is 8 inches (24/3). Using Corollary 7.5 to find the area of the triangle:

$$A = \frac{8^2\sqrt{3}}{4} = \frac{64\sqrt{3}}{4} = 16\sqrt{3} \text{ in}^2.$$

37. $AG = CG = 1$ inch since $ABCG$ is a square. The area of $\triangle ACG$ $\dfrac{1}{2}(1)(1) = \dfrac{1}{2}$ in^2. $CEFG$ is a trapezoid where $\overline{CG}$ and $\overline{EF}$ are the parallel bases and $\overline{FG}$ is the height. The area of the trapezoid is $\dfrac{1}{2}(4+1)4 = 10$ in^2. The area of the shaded region is $\dfrac{1}{2} + 10 = 10.5$ or $\dfrac{21}{2}$ in^2.

CHAPTER 8 SOLID GEOMETRY

Section 8.1 Planes and Polyhedrons

1. True

3. False (actually $P \perp R$)

5. True

7. False

9. True

11. True

13. $f = 7, v = 6, e = 11$

$7 + 6 - 11 = 2$

15. $f = 6, v = 6, e = 10$

$6 + 6 - 10 = 2$

17. $f = 8, v = 6, e = 12$

$8 + 6 - 12 = 2$

19. Since $v = 12$ and $e = 30$, substitute into $f + v - e = 2$ to find f.

$$f + 12 - 30 = 2$$
$$f - 18 = 2$$
$$f = 20$$

The polyhedron has 20 faces.

21.

23.

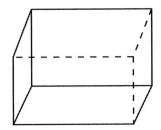

25.

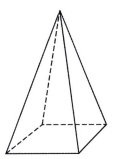

27. A cube has 6 faces and each face is a square. The area of each face is $12(12) = 144$ ft^2. Total surface area of the cube is $6(144) = 864$ ft^2. Since each square foot in insulating material costs \$2.50 per square foot, the total cost is $(864)(2.50) = \$2160.00$.

29. Consider the cube with edge e shown below.

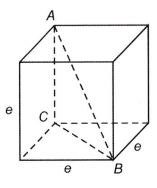

Consider the diagonal of the cube $\overline{AB}$ and the diagonal of its base $\overline{BC}$. By the Pythagorean Theorem, $(BC)^2 = e^2 + e^2 = 2e^2$, so $BC = \sqrt{2e^2} = e\sqrt{2}$. Then $\triangle ABC$ is a right triangle with hypotenuse $\overline{AB}$ and legs of length e and $e\sqrt{2}$. Using the Pythagorean Theorem again, $(AB)^2 = e^2 + \left(e\sqrt{2}\right)^2 = e^2 + 2e^2 = 3e^2$ so that $AB = \sqrt{3e^2} = e\sqrt{3}$. Thus, the length of each diagonal is $e\sqrt{3}$.

31. Answers will vary.

Section 8.2 Prisms

8.2 PRACTICE EXERCISES

1. The lateral area of the workshop is needed.
   $(LA = ph)$. The perimeter of the room is
   $16 + 18 + 16 + 18 = 68$.

   $$LA = 68(8) = 544 \text{ ft}^2$$

   Since the siding costs $1.50 per square foot,
   the cost of the project is $(544)(1.50) =$
   $816.00.

2. The volume of the truck bed is needed.
   $(V = Bh)$

   $$B = (5.5)(2.5) = 13.75$$
   $$V = (13.75)(2.0) = 27.5 \text{ yd}^3$$

8.2 SECTION EXERCISES

1. True

   $$SA = LA + 2B$$
   $$= ph + 2lw \text{ since } B = lw$$
   $$= (2l + 2w)h + 2lw \text{ since } p = 2l + 2w$$
   $$= 2lh + 2wh + 2lw \text{ (distributive property)}$$
   $$= 2lw + 2lh + 2wh$$

3. True

5. For this cube, $l = w = h = 1.5$.

   The formula for the total surface area of a
   right prism is $SA = LA + 2B$ where $LA = ph$
   and $B = lw$.

   $$p = 4(1.5) = 6$$
   $$SA = 6(1.5) + 2(1.5)^2 = 13.5 \text{ m}^2$$

7. The formula for the total surface area of a
   right prism is $SA = LA + 2B$ where $LA = ph$
   and $B = lw$.

   $$p = 2(6.4) + 2(15.5) = 43.5$$
   $$SA = 43.8(20.2) + 2(6.4)(15.5)$$
   $$= 1083.16 \text{ cm}^2$$

9. The formula for the total surface area of a
   right prism is $SA = LA + 2B$ where $LA = ph$
   and $B = \frac{1}{2}bh$.

   To find the perimeter of the base, use the
   Pythagorean Theorem. See the figure.

   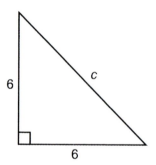

   $$6^2 + 6^2 = c^2$$
   $$72 = c^2$$
   $$c = \sqrt{72}$$
   $$p = 6 + 6 + \sqrt{72} = 12 + \sqrt{72}$$
   $$SA = \left(12 + \sqrt{72}\right)(11) + (2)\left(\frac{1}{2}\right)(6)(6)$$
   $$= 132 + 11\sqrt{72} + 36$$
   $$= \left(168 + 11\sqrt{72}\right) \text{ cm}^2 \text{ or}$$
   $$\left(168 + 66\sqrt{2}\right) \text{ cm}^2$$

11. The formula for the total surface area of a right prism is $SA = LA + 2B$ where $LA = ph$ and $\frac{1}{2}bh$.

Find the second leg of the right triangle using the Pythagorean Theorem.

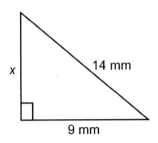

$$x^2 + 9^2 = 14^2$$
$$x^2 + 81 = 196$$
$$x^2 = 115$$
$$x = \sqrt{115}$$

The perimeter of the triangle is $p = 9 + 14 + \sqrt{115} = 23 + \sqrt{115}$. Note: The base and height of a triangle are perpendicular.

$$SA = \left(23 + \sqrt{115}\right)(22) + 2\left(\frac{1}{2}\right)(9)\left(\sqrt{115}\right)$$
$$= 506 + 22\sqrt{115} + 9\sqrt{115}$$
$$= \left(506 + 31\sqrt{115}\right) \text{ mm}^2$$

13. The formula for the volume of a right prism is $V = Bh$. In this problem $l = w = h = 10.2$.

$$B = (10.2)^2 = 104.04$$
$$V = 104.04(10.2) = 1061.208 \text{ cm}^3$$

15. The formula for the volume of a right prism is $V = Bh$.

$$B = lw = (3.4)(9.5) = 32.3$$
$$V = 32.3(8.75) = 282.625 \text{ in}^3$$

17. The formula for the volume of a right prism is $V = Bh$.
Since the base is an equilateral triangle, use the formula $B = \dfrac{a^2\sqrt{3}}{4}$ where a = side of equilateral triangle, to find the area.

$$B = \frac{8^2\sqrt{3}}{4} = 16\sqrt{3}$$
$$V = \left(16\sqrt{3}\right)(24) = 384\sqrt{3} \text{ cm}^3$$

19. The formula for the volume of a right prism is $V = Bh$.
Since the base is a regular hexagon, use the formula $B = \frac{1}{2}ap$, where a is the apothem, to find the area.
The perimeter of the hexagon is $6(3.6) = 21.6$.
The apothem can be found by using the figure shown and the 30°-60°-90° Triangle Theorem.

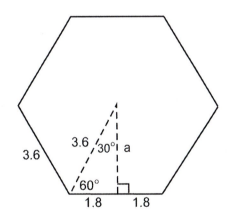

$$a = \frac{\sqrt{3}}{2}(3.6) = 1.8\sqrt{3}$$
$$B = \frac{1}{2}(21.6)\left(1.8\sqrt{3}\right) = 19.44\sqrt{3}$$
$$V = \left(19.44\sqrt{3}\right)(4.8) = 93.312\sqrt{3} \text{ yd}^3$$

21. The formula for the total surface area of a cube is $SA = 6s^2$ (using the ideas developed in the *Technology Connection*) where S = length of one edge of the cube.

$$SA = 6s^2$$
$$6s^2 = 294$$
$$s^2 = 49$$
$$s = 7$$

The length of one edge is 7 cm.

The formula for the volume of a cube is $V = s^3$.

$$V = 7^3 = 343 \text{ cm}^3$$

An alternative solution using Theorems 8.1 and 8.2:

The formula for the total surface area of a right prism is $SA = LA + 2B$ where $LA = ph$ and $B = x^2$ where x = length of one edge of the cube. The perimeter of the cube is $4x$.

$$LA = 4x(x) = 4x^2$$
$$SA = 4x^2 + 2x^2 = 6x^2$$
$$6x^2 = 294$$
$$x^2 = 49$$
$$x = 7$$

The length of one edge is 7 cm.
The formula for the volume of a right prism is $V = Bh$.
The base is a square so the area is $x^2 = 7^2 = 49$.

$$V = 49(7) = 343 \text{ cm}^3$$

23. Find the volume of the "outer" prism and subtract the volume of the "inner" prism. The formula for the volume of a right prism is $V = Bh$.

Volume of Outer Prism

$$B = (7)(12) = 84$$
$$V = (84)(20) = 1680$$

Volume of Inner Prism

$$B = (2)(3) = 6$$
$$V = (6)(20) = 120$$

The volume of the given prism is $1680 - 120 = 1560 \text{ ft}^3$.

25. First find the volume of the cube using the formula $V = s^3$ from *Technology Connections*.

$$V = 22^3 = 10648 \text{ cm}^3$$

Since each cubic centimeter of water weighs 1 gram, the weight of the water is 10,648 g.

27. First, find the lateral area of the prism (siding only goes on the sides of the building) $LA = ph$

$$p = 2(30) + 2(12) = 84$$
$$LA = 84(8) = 672 \text{ ft}^2$$

Since siding costs \$0.75 per square foot, the cost of the project is $672(0.75) = \$504.00$

29. Consider the right prism shown in the figure below.

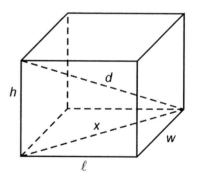

To find d we first find x using the Pythagorean Theorem.

$$x^2 = \ell^2 + w^2$$
$$x = \sqrt{\ell^2 + w^2}$$

Then by using the Pythagorean Theorem again,

$$d^2 = x^2 + h^2 = \ell^2 + w^2 + h^2$$

$$d = \sqrt{\ell^2 + w^2 + h^2}.$$

31. Answers will vary.

Section 8.3 Pyramids

8.3 PRACTICE EXERCISES

1. The formula for the total surface area of a right pyramid is $SA = LA + B$ where $LA = \frac{1}{2}pl$. Since each face is an equilateral triangle the base is a square, with an edge measuring $3\sqrt{3}$ meters long. The area of the base is $A = \left(3\sqrt{3}\right)^2 = 27$ m^2. The perimeter of the base is $4\left(3\sqrt{3}\right) = 12\sqrt{3}$.

$$LA = \frac{1}{2}\left(12\sqrt{3}\right)(4.5) = 27\sqrt{3}$$

$$SA = 27\sqrt{3} + 27$$

Thus, to the nearest tenth, the total surface area is approximately 73.8 m^2.

2. The formula for the volume of a right pyramid is $V = \frac{1}{3}Bh$.

Since each face is an equilateral triangle, the base is a square with an edge measuring $3\sqrt{3}$ meters long. The area of the base is

$$A = \left(3\sqrt{3}\right)^2 = 27 \text{ m}^2.$$

$$V = \frac{1}{3}(27)(3.7) = 33.3 \text{ m}^2$$

Thus, the volume of the pyramid is 33.3 m^2.

8.3 SECTION EXERCISES

1. The formula for the lateral surface area of a right pyramid is $LA = \frac{1}{2}pl$.

The perimeter of the equilateral triangular base is $p = 3(22) = 66$.

$$p = 3(22) = 66$$

$$LA = \frac{1}{2}pl = \frac{1}{2}(66)(28) = 924 \text{ m}^2$$

Thus, the lateral area is 924 m^2.

3. The formula for the lateral surface area of a right pyramid is $LA = \frac{1}{2}pl$. The perimeter of the square base is $p = 4(124) = 496$. To find the slant height, use the Pythagorean Theorem.

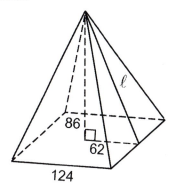

$$86^2 + 62^2 = \ell^2$$

$$11240 = \ell^2$$

$$\ell = \sqrt{11240}$$

$$LA = \frac{1}{2}p\ell = \frac{1}{2}(496)\left(\sqrt{11,240}\right)$$

Thus the lateral area of the pyramid is $248\sqrt{11,240}$ cm^2.

5. The formula for the surface area of a right pyramid is $SA = LA + B$ where $LA = \frac{1}{2}p\ell$ and $B = s^2 = 4^2 = 16$.

$$p = 4(4) = 16$$

To find the slant height (ℓ), use the Pythagorean Theorem.

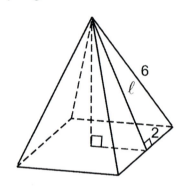

$$2^2 + \ell^2 = 6^2$$

$$4 + \ell^2 = 36$$

$$\ell^2 = 32$$

$$\ell = \sqrt{32}$$

$$LA = \frac{1}{2}p\ell = \frac{1}{2}(16)\left(\sqrt{32}\right) = 8\sqrt{32}$$

$$SA = LA + B = 8\sqrt{32} + 16$$

Thus the approximate surface area, rounded to the nearest tenth, is 61.3 ft^2. Remember, only round the final answer.

7. The formula for the surface area of a right pyramid is $SA = LA + B$ where $LA = \frac{1}{2}p\ell$ and $B = \frac{1}{2}ap$.

Since the base is a regular hexagon, the perimeter is $p = 6(7) = 42$. To find the apothem, see the figure of the regular hexagon and use the 30°-60°-90° Triangle Theorem.

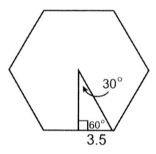

Recall the length of the long leg is $\sqrt{3}$ times the short leg. Since the apothem is the long leg, $a = 3.5\sqrt{3}$.

$$SA = LA + B = \frac{1}{2}p\ell + \frac{1}{2}ap$$

$$= \frac{1}{2}(42)(24.8) + \frac{1}{2}\left(3.5\sqrt{3}\right)(42)$$

$$= 520.8 + 73.5\sqrt{3}$$

Thus the surface area of the pyramid, rounded to the nearest tenth, is approximately 648.1 mm^2.

9. The formula for the volume of a regular right pyramid is $V = \frac{1}{3}Bh$. Since the base is a square, the area is $B = 22^2 = 484$.

$$V = \frac{1}{3}(484)(15) = 2420$$

Thus the volume of the pyramid is 2420 yd^3.

11. The formula for the volume of a regular right pyramid is $V = \frac{1}{3}Bh$. Since the base is a regular hexagon, $B = \frac{1}{2}ap$. The perimeter is $p = 6(16) = 96$. To find the apothem, see the figures of a regular hexagon and use the 30°-60°-90° Triangle Theorem.

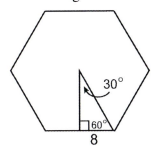

Recall the length of the long leg is $\sqrt{3}$ times the short leg. Since the apothem is the long leg, $a = 8\sqrt{3}$.

$$B = \frac{1}{2}\left(8\sqrt{3}\right)(96) = 384\sqrt{3}$$

To find the height of the pyramid, use the Pythagorean Theorem knowing the slant height is 24 and the apothem is $8\sqrt{3}$.

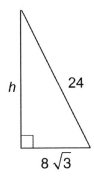

$$h^2 + \left(8\sqrt{3}\right)^2 = 24^2$$

$$h^2 + 192 = 576$$

$$h^2 = 384$$

$$h = \sqrt{384}$$

$$V = \frac{1}{3}\left(384\sqrt{3}\right)\left(\sqrt{384}\right)$$

Thus the volume of the pyramid, rounded to the nearest tenth, is approximately 4344.5 cm².

13. The formula for the volume of a regular right pyramid is $V = \frac{1}{3}Bh$.
In problem 5, $B = 4^2 = 16$.

$$V = \frac{1}{3}(16)\left(2\sqrt{7}\right)$$

Thus the volume of the pyramid, rounded to the nearest tenth, is approximately 28.2 ft³.

15. The formula for the volume of a regular right pyramid is $V = \frac{1}{3}Bh$. In problem 7,

$$B = \frac{1}{2}\left(3.5\sqrt{3}\right)(42) = 73.5\sqrt{3}.$$

$$V = \frac{1}{3}\left(73.5\sqrt{3}\right)(24.0) = 588\sqrt{3}$$

Thus the volume of the pyramid, rounded to the nearest tenth, is approximately 1018.4 mm³.

17. The formula for the lateral area of a pyramid is $LA = \frac{1}{2}p\ell$. The perimeter of the square base is $4(100) = 400$. Use the Pythagorean Theorem to find the slant height.

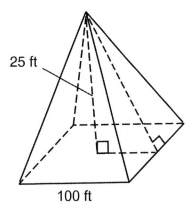

$$(25)^2 + (50)^2 = \ell^2$$

$$625 + 2500 = \ell^2$$

$$3125 = \ell^2$$

$$\ell = \sqrt{3125}$$

$$LA = \frac{1}{2}p\ell = \frac{1}{2}(400)\left(\sqrt{3125}\right)$$

$$= 200\sqrt{3125}$$

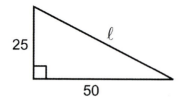

The lateral area of the pyramid, rounded to the nearest tenth, is approximately 11,180.3 ft². One gallon of paint covers 220 ft², so to figure the number of gallons needed, divide the lateral area by 220.

$$\frac{11180.3}{220} \approx 50.8$$

Thus it takes about 50.8 or about 51 gallons of paint for the project.

Section 8.4 Cylinders and Cones

8.4 PRACTICE EXERCISES

1. To find the surface area use the formula $SA = LA + 2B$ or $2\pi rh + 2\pi r^2$.

$$SA = 2\pi(12.8)(22.3) + 2\pi(12.8)^2$$

Thus, the surface area of the cylinder, rounded to the nearest tenth, is approximately 2822.9 cm².

Remember, when the final answer is to be approximated, be sure to only round the final answer not the calculations needed to find the final answer.

2. First find the volume of the storage tank using the formula $V = Bh$ or $\pi r^2 h$.

$$V = \pi(6)^2 (10) = 360\pi \text{ m}^3$$

Now find the volume of each water truck:

$$V = \pi(0.8)^2 (5) = 3.2\pi \text{ m}^3.$$

19. Use the volume formula to find the amount of wax needed, $V = \frac{1}{3} Bh$. The base is a square so the area is $B = 10^2 = 100$.

$$V = \frac{1}{3}(100)(15) = 500 \text{ cm}^3$$

Thus, 500 cm³ of wax is needed.

21. There are 4 faces on the tetrahedron. They all appear congruent.

To find the number of trucks that could be filled from the tank, divide the volume of the tank by the volume of one truck.

$$\frac{360\pi}{3.2\pi} = 112.5$$

Thus 113 trucks will be needed to remove all the water. (There would not be $\frac{1}{2}$ a truck.)

3. Use the formula for the lateral area of a cone, $LA = \frac{1}{2} p\ell$ or $\pi r\ell$. Use the Pythagorean Theorem to find the slant height (ℓ).

$$(7.4)^2 + (16.2)^2 = \ell^2$$
$$54.76 + 262.44 = \ell^2$$
$$317.2 = \ell^2$$
$$\ell = \sqrt{317.2}$$
$$LA = \pi(7.4)\left(\sqrt{317.2}\right)$$

Thus the lateral area, rounded to the nearest tenth, is approximately 414.0 ft².

8.4 SECTION EXERCISES

1. True

3. The formula for the surface area of a cylinder is $SA = LA + 2B$ or $2\pi rh + 2\pi r^2$.

$$SA = 2\pi(6)(5) + 2\pi(6)^2 = 132\pi$$

Thus, the surface area of the cylinder, rounded to the nearest tenth, is approximately 414.7 cm^2.

5. The formula for the surface area of a cylinder is $SA = LA + 2B$ or $2\pi rh + 2\pi r^2$.

$$SA = 2\pi(12.2)(30.0) + 2\pi(12.2)^2$$
$$= 1029.68\pi$$

Thus, the surface area of the cylinder, rounded to the nearest tenth, is approximately 3234.8 in^2.

7. The formula for the surface area of a cylinder is $SA = LA + 2B$ or $2\pi rh + 2\pi r^2$.

Since $d = 11.4$, $r = \dfrac{11.4}{2} = 5.7$.

$$SA = 2\pi(5.7)(4.4) + 2\pi(5.7)^2$$
$$= 115.14\pi$$

Thus, the surface area of the cylinder, rounded to the nearest tenth, is approximately 361.7 m^2.

9. The formula for the volume of a cylinder is $V = Bh$ or $\pi r^2 h$.

$$V = \pi(7)^2(8) = 392\pi$$

Thus, the volume of the cylinder, rounded to the nearest tenth, is approximately 1231.5 cm^3.

11. The formula for the volume of a cylinder is $V = Bh$ or $\pi r^2 h$.

$$V = \pi(14.5)^2(35.5) = 7463.875\pi$$

Thus, the volume of the cylinder, rounded to the nearest tenth, is approximately 23,448.5 in^3.

13. The formula for the volume of a cylinder is $V = Bh$ or $\pi r^2 h$.

Since $d = 12.6$, $r = \dfrac{12.6}{2} = 6.3$.

$$V = \pi(6.3)^2(16.2) = 642.978\pi$$

Thus, the volume of the cylinder, rounded to the nearest tenth, is approximately 2020.0 m^3.

15. The formula for the surface area of a cone is $SA = LA + B$ or $\pi rl + \pi r^2$.

$$SA = \pi(3)(5) + \pi(3)^2 = 24\pi$$

Thus, the surface area of the cone, rounded to the nearest tenth, is approximately 75.4 yd^2.

17. The formula for the surface area of a cone is $SA = LA + B$ or $\pi r\ell + \pi r^2$.

$$SA = \pi(28.4)(42.6) + \pi(28.4)^2$$
$$= 2016.4\pi$$

Thus, the surface area of the cone, rounded to the nearest tenth, is approximately 6334.7 in^2.

19. The formula for the surface area of a cone is $SA = LA + B$ or $\pi r\ell + \pi r^2$. Use the Pythagorean Theorem to find the slant height.

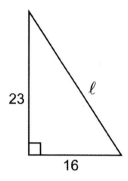

$$16^2 + 23^2 = \ell^2$$
$$256 + 529 = \ell^2$$
$$785 = \ell^2$$
$$\ell = \sqrt{785}$$
$$SA = \pi(16)\left(\sqrt{785}\right) + \pi(16)^2$$

Thus, the surface area of the cone, rounded to the nearest tenth, is approximately 2212.6 ft^2.

21. The formula for the volume of a cone is $V = \frac{1}{3}Bh$ or $\frac{1}{3}\pi r^2 h$.

$$V = \frac{1}{3}\pi(3)^2(4) = 12\pi$$

Thus, the volume of the cone, rounded to the nearest tenth, is approximately 37.7 in^3.

23. The formula for the volume of a cone is $V = \frac{1}{3}Bh$ or $\frac{1}{3}\pi r^2 h$.

$$V = \frac{1}{3}\pi(6.9)^2(14.2) = 225.354\pi$$

Thus, the volume of the cone, rounded to the nearest tenth, is approximately 708.0 cm^3.

25. The formula for the volume of a cone is $V = \frac{1}{3}Bh$ or $\frac{1}{3}\pi r^2 h$.

Use the Pythagorean Theorem to find the height (h) of the cone.

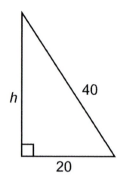

$$20^2 + h^2 = 40^2$$
$$400 + h^2 = 1600$$
$$h^2 = 1200$$
$$h = \sqrt{1200}$$
$$V = \frac{1}{3}\pi(20)^2\left(\sqrt{1200}\right)$$

Thus, the volume of the cone, rounded to the nearest tenth, is approximately 14,510.4 m^3.

27. The formula for the volume of a cylinder is $V = Bh$ or $\pi r^2 h$.

Since the radius is the same but the height is doubled, in the formula, use $2h$ instead of h.

$$V = \pi r^2(2h) = 2\left(\pi r^2 h\right)$$

The volume of the cylinder is doubled the original volume: 2(26.8) = 53.6.

The volume of the cylinder with double the original height is 53.6 cm^3.

If the height is the same but the radius is twice as long, in the formula, use $2r$ instead of r.

$$V = \pi(2r)^2 h = 4\left(\pi r^2 h\right)$$

The volume of the cylinder multiplied by a factor of 4: 4(26.8) = 107.2 cm^3.

29. If the prism is inscribed within the cylinder, then the square base of the prism is inscribed in the circular base of the cylinder as shown in the figure.

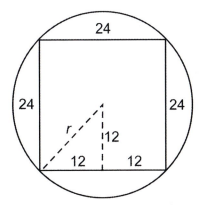

The radius of the cylinder is found using the 45°-45°-90° triangle with legs 12 inches. The hypotenuse r is $r = 12\sqrt{2}$.

The height of the cylinder is the same as the height of the prism, 16 inches. Then

$$SA = 2\pi rh + 2\pi r^2$$
$$= 2(\pi)(12\sqrt{2})(16) + 2\pi(12\sqrt{2})^2$$

Thus, the surface area of the cylinder is approximately 3515.6 in² correct to the nearest tenth.

$$V = \pi r^2 h$$
$$= (\pi)(12\sqrt{2})^2(16)$$

Thus, the volume is approximately 14,476.5 in³, correct to the nearest tenth.

31. First, find the volume of the pie pan. The formula for the volume of a cylinder is $V = Bh$ or $\pi r^2 h$. Since the diameter is 20, the radius is 10 cm.

$$V = \pi(10)^2(3) = 300\pi$$

Next, find the volume of the can of pie filling. Since the diameter is 8, the radius is $\frac{8}{2} = 4$ cm.

$$V = \pi(4)^2(12) = 192\pi \text{ cm}^3$$

To find the number of cans of pie filling needed

$$\frac{\text{volume of pie pan}}{\text{volume of can of filling}} = \frac{300\pi \text{ cm}^3}{192\pi \text{ cm}^3}$$

Thus, it takes approximately 1.6 cans of cherry pie filling to fill the pan.

33. The surface area of the tank is needed; use the formula $SA = LA + 2B$ or $2\pi rh + 2\pi r^2$.

$$SA = 2\pi(3.8)(9.8) + 2\pi(3.8)^2$$
$$= 103.36\pi$$

To find the number of gallons needed, divide $\frac{103.36\pi}{150}$.

Since one gallon of paint covers 150 ft², approximately 2.2 gallons of paint is needed.

35. The lateral area of the cone is needed. Use the formula $LA = \frac{1}{2}p\ell$ or $\pi r\ell$.

Use the Pythagorean Theorem to find the slant height (ℓ).

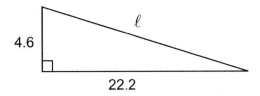

$$(4.6)^2 + (22.2)^2 = \ell^2$$
$$21.16 + 492.84 = \ell^2$$
$$514 = \ell^2$$
$$\ell = \sqrt{514}$$
$$LA = \pi(4.6)(\sqrt{514})$$
$$LA = \pi(4.6)(\sqrt{514})$$

To find the number of gallons needed, divide $\frac{\pi(4.5)(\sqrt{514})}{180}$.

Since one gallon of paint covers 180 ft², approximately 1.8 gallons of paint is needed.

37. First, find the volume of the cylindrical container using the formula $V = Bh$ or $\pi r^2 h$. Since the diameter is 5 inches, the radius is $\frac{5}{2} = 2.5$ in.

$$V = \pi(2.5)^2(9.5) = 59.375\pi$$

The volume of the cylinder, rounded to the nearest tenth, is approximately 186.5 in³. Next, find the volume of the rectangular prism using the formula $V = Bh$ where $B = (2.25)(6.5) = 14.625$.

$$V = (14.625)(11) = 160.875 \text{ in}^3$$

Thus the cylinder holds more cereal.

39. The lateral area of a cylinder (bird feeder) is needed. Use the formula $LA = ph$ or $2\pi rh$.

Since the diameter is 2 inches, the radius is $\frac{2}{2} = 1$ inch.

$$LA = 2\pi(1)(18) = 36\pi.$$

Jamal will need approximately, rounded to the nearest tenth, 113.1 in² of fencing.

Section 8.5 Spheres and Composite Figures

8.5 PRACTICE EXERCISE

1. The surface area of the tank is needed. Use the formula $SA = 4\pi r^2$.

$$SA = 4\pi(3.2)^2 = 40.96\pi$$

Since each square meter of material costs \$2.50, the total cost of the project is $(40.96\pi)(2.50)$. Rounding the answer to the nearest cent, the cost is about \$321.70.

8.5 SECTION EXERCISES

1. The formula for the surface area of a sphere is $SA = 4\pi r^2$.

$$SA = 4\pi(12)^2 = 576\pi$$

Thus, the surface area of the sphere, rounded to the nearest tenth, is approximately 1809.6 m².

3. The formula for the surface area of a sphere is $SA = 4\pi r^2$.

$$SA = 4\pi(0.75)^2 = 2.25\pi$$

Thus, the surface area of the sphere, rounded to the nearest tenth, is approximately 7.1 ft².

5. The formula for the surface area of a sphere is $SA = 4\pi r^2$.

$$SA = 4\pi(5.6)^2 = 125.44\pi$$

Thus, the surface area of the sphere, rounded to the nearest tenth, is approximately 394.1 yd².

7. The formula for the volume of a sphere is $V = \frac{4}{3}\pi r^3$.

$$V = \frac{4}{3}\pi(7)^3$$

Thus, the volume of the sphere, rounded to the nearest tenth, is approximately 1436.8 ft³.

9. The formula for the volume of a sphere is
$V = \frac{4}{3}\pi r^3$.

$$V = \frac{4}{3}\pi(0.90)^3$$

Thus, the volume of the sphere, rounded to the nearest tenth, is approximately 3.1 m^3.

11. The formula for the volume of a sphere is
$V = \frac{4}{3}\pi r^3$.

$$V = \frac{4}{3}\pi(13.8)^3$$

Thus, the volume of the sphere, rounded to the nearest tenth, is approximately 11008.4 yd^3.

13. From the *Technology Connection* in Section 8.2, the formula for the surface area of a cube is $SA = 6s^2$. The surface area of the cube is 150 in^2.

$$150 = 6s^2$$
$$25 = s^2$$
$$s = 5$$

The volume of a cube with a side of 5 inches is $V = s^3 = 5^3 = 125$ in^3.

The formula for the surface area of a sphere is $SA = 4\pi r^2$.
The surface area of this sphere is 150 in^2.

$$150 = 4\pi r^2$$
$$\frac{150}{4\pi} = r^2$$
$$r = \sqrt{\frac{150}{4\pi}}$$

The volume of a sphere with this radius is

$$V = \frac{4}{3}\pi\left(\sqrt{\frac{150}{4\pi}}\right)^3.$$

The volume of the sphere, rounded to the nearest tenth, is approximately 172.7 in^3. Thus, the sphere has the greater volume.

15. The formula for the surface area of a sphere is $SA = 4\pi r^2$. The formula for the volume of a sphere is $V = \frac{4}{3}\pi r^3$.

$$4\pi r^2 = \frac{4}{3}\pi r^3$$

$$r^2 = \frac{1}{3}r^3 \quad \text{(Dividing both sides by } 4\pi)$$

$$3r^2 = r^3 \quad \text{(Multiply both sides by 3)}$$

$$0 = r^3 - 3r^2$$

$$0 = r^2(r-3)$$

$$r^2 = 0 \quad \text{or} \quad r - 3 = 0$$

$$r = 0 \quad \text{or} \quad r = 3$$

Since $r = 0$ is meaningless (a sphere cannot have a radius of 0), the radius of the sphere with the same volume and surface area is 3 units.

17. The volume formula for a sphere is needed,

$$V = \frac{4}{3}\pi r^3.$$

The volume for one apple with a 4 inch diameter (2 inch radius) is

$$V = \frac{4}{3}\pi(2)^3 = \frac{32\pi}{3}.$$

Since the recipe calls for 5 apples with a 4 inch diameter, the total volume of apples needed is

$$5\left(\frac{32\pi}{3}\right) = \frac{160\pi}{3}.$$

The volume of an apple with a diameter of 3 inches (radius = 1.5 inches) is

$$V = \frac{4}{3}\pi(1.5)^3 = 4.5\pi.$$

To determine how many 3 inch apples are needed, divide the volume needed for the pie by the volume of one 3 inch apple.

$$\frac{160\pi}{3} \div 4.5\pi$$

Thus, rounded to the nearest whole apple, 12 of the smaller apples are needed.

19. The formula for the surface area of a hemisphere is one half the surface area of a sphere.

$$SA_{hemisphere} = \frac{1}{2}\left(4\pi r^2\right)$$

Since the diameter is 14 inches, the radius is 7 inches.

$$SA_{hemisphere} = \frac{1}{2}\left(4\pi\right)\left(7^2\right) = 98\pi$$

Thus, the surface area of the pot, rounded to the nearest tenth, is approximately 307.9 in^2.

21. The formula for the volume of a sphere is needed, $V = \frac{4}{3}\pi r^3$.

$$V = \frac{4}{3}\pi\left(32\right)^3$$

Rounded to the nearest tenth, the volume of the tank is approximately 137,258.3 ft^3.

23. To find the amount of glue in the bottle, add the volume of the cylinder to the volume of the cone. The formula for the volume of a cylinder is $V = Bh$ or $\pi r^2 h$.

$$V_{cylinder} = \pi\left(3\right)^2\left(10\right) = 90\pi$$

The formula for the volume of the cone is $V = \frac{1}{3}Bh$ or $\frac{1}{3}\pi r^2 h$.

The radius of the cone is the same as the radius of the cylinder, $r = 3$ cm.

Use the Pythagorean Theorem to find the height (h) of the cone.

$$3^2 + h^2 = 5^2$$
$$9 + h^2 = 25$$
$$h^2 = 16$$
$$h = 4$$
$$V_{cone} = \frac{1}{3}\pi\left(3\right)^2\left(4\right) = 12\pi$$

Volume of the glue bottle is

$$V_{cylinder} + V_{cone} = 90\pi + 12\pi = 102\pi.$$

Thus, rounded to the nearest tenth, the glue bottle will hold approximately 320.4 cm^3 of glue.

25. The surface area of the chemical storage tank will be the surface areas of the two hemispheres and the lateral area of the cylindrical tank. The formula for the lateral area of a cylinder is $LA = ph$ or $2\pi rh$.

$$LA_{cylinder} = 2\pi\left(2.8\right)\left(16.2\right) = 90.72\pi$$

The surface area of the two hemispheres is equal to the surface area of one sphere with $r = 2.8$.

The formula for the surface area of a sphere is $SA = 4\pi r^2$.

$$SA_{sphere} = 4\pi\left(2.8\right)^2 = 31.36\pi$$

The surface area to be insulated is $90.72\pi + 31.36\pi = 122.08\pi$.

To find the cost, multiply the surface area by 1.25: cost $= (122.08\pi)(1.25)$.

Thus the cost, rounded to the nearest cent, is approximately $479.41.

27. The volume of the cone and the hemisphere are needed. Since the overall height of the cone and the hemisphere is 6.4 inches, subtract the radius of the hemisphere (1.2 inches) to find the height (h) of the cone.

$$h = 6.4 - 1.2 = 5.2 \text{ inches}$$

The formula for the volume of the cone is $V = \frac{1}{3}Bh$ or $\frac{1}{3}\pi r^2 h$.

$$V_{cone} = \frac{1}{3}\pi(1.2)^2(5.2) = 2.496\pi$$

The volume of a hemisphere is one-half the volume of a sphere.

$$V_{hemisphere} = \frac{1}{2} \cdot \frac{4}{3}\pi r^3$$
$$= \frac{1}{2} \cdot \frac{4}{3}\pi(1.2)^3 = 1.152\pi$$

The total volume of ice cream is $2.496\pi + 1.152\pi = 3.648\pi$.

The total volume of ice cream, rounded to the nearest tenth, is approximately $11.5\ in^3$.

29. The surface area of the machine part is needed which is
$$SA_{prism} - 2(A_{circles}) + LA_{cylinder}.$$

The formula for the surface area of a right prism is $SA = LA + 2B$ where $LA = ph$.

The perimeter of the prism is $p = 2(6.4) + 2(5.8) = 24.4$. The area of the base is $B = 6.4(5.8) = 37.12$.

$$SA_{prism} = (24.4)(2.3) + 2(37.12)$$
$$= 130.36$$

The formula for the area of a circle is $A = \pi r^2$.

$$A_{circle} = \pi(1.8)^2 = 3.24\pi$$

The area of two circles is needed: $2(3.24\pi) = 6.48\pi$.

The formula for the lateral area of a cylinder is $LA = ph$ or $2\pi rh$.

$$LA_{cylinder} = 2\pi(1.8)(2.3) = 8.28\pi$$

The surface area of the machine part is

$$SA_{prism} - 2(A_{circles}) + LA_{cylinder}$$
$$= 130.36 - 6.48\pi + 8.28\pi$$

The cost of the finishing operation is $(130.36 - 6.48\pi + 8.28\pi)(0.26)$

Thus, the cost of the finishing operations, rounded to the nearest cent, is approximately $35.36.

Chapter 8 Review Exercises

1. True   **2.** True

3. True   **4.** True

5. False ($\mathcal{L}$ and $\mathcal{R}$ are perpendicular)

6. False (u and $\mathcal{P}$ are perpendicular)

7. $f = 6$, $y = 8$, $e = 12$;
$6 + 8 - 12 = 2$

8. $f = 12$, $y = 8$, $e = 18$;
$12 + 8 - 18 = 2$

9. Use $f + v - e = 2$ where $v = 14$ and $e = 36$. Solve for f.
$$f + 14 - 36 = 2$$
$$f - 22 = 2$$
$$f = 24$$

The polyhedron has 24 faces.

10. A regular octahedron has a total of 12 edges. If each edge is 2.5 ft in length, the total length of the tubing is $(12)(2.5) = 30$ ft. Since each foot of tubing costs $1.65, the total cost of the project is $(30)(1.65) = $49.50.

11. The formula for the surface area of a right prism is $SA = LA + 2B$ where $LA = ph$. Since the base is an equilateral triangle the perimeter is $p = 3(12.6) = 37.8$.

The area of one base is the area of an equilateral triangle,

$$A = \frac{a^2\sqrt{3}}{4} = \frac{(12.6)^2\sqrt{3}}{4}$$
$$= 39.69\sqrt{3}.$$

$$SA = 37.8(21.3) + 2\left(39.69\sqrt{3}\right)$$

Thus, the surface area of the prism, rounded to the nearest tenth, is approximately 942.6 cm^2.

12. The formula for the surface area of a right prism is $SA = LA + 2B$ where $LA = ph$. Since the base is a rectangle, the perimeter is $p = 2(9.6) + 2(5.8) = 30.8$.

The formula for the area of a base is

$$A = lw = (9.6)(5.8) = 55.68.$$
$$SA = (30.8)(2.6) + 2(55.68)$$
$$= 191.44$$

Thus the surface area of the prism is 191.44 ft^2.

(Note: because the calculations were exact to two decimal places, the answer was not rounded.)

13. The formula for the volume of a cube is $V = s^3$, from *Technology Connections* in Section 8.2.

$$V = (18.65)^3 = 6434.856$$

Thus the volume of the cube is exactly 6434.856 m^3.

14. The volume of this right prism is given by $V = Bh$ where B is the area of its base and $h = 7.4$ m is its height. To find B we must find the area of the regular hexagon shown below.

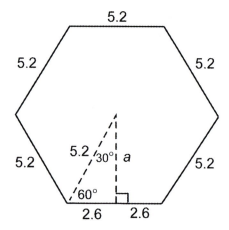

$B = \frac{1}{2}ap$ where a is the apothem and p is the perimeter of the base. Then $a = 2.6\sqrt{3}$ by properties of a 30°-60°-90° triangle and $p = 6(5.2) = 31.2$. Thus,

$$B = \frac{1}{2}\left(2.6\sqrt{3}\right)(31.2)$$

so that

$$V = Bh = \frac{1}{2}\left(2.6\sqrt{3}\right)(31.2)(7.4)$$
$$= 300.144\sqrt{3}.$$

The volume is about 519.9 m^3, correct to the nearest tenth of a cubic meter.

15. The formula for the volume of a right prism is needed, $V = Bh$.

The base is a regular hexagon and the formula for the area of a regular figure is $A = \frac{1}{2}ap$.

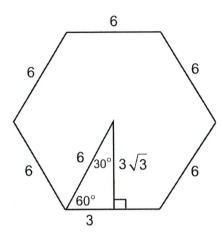

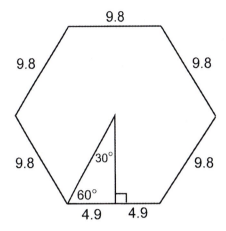

Each side of the hexagon is 6 ft so the perimeter is 6(6) = 36. To find the apothem, use the 30°-60°-90° Triangle Theorem.

The apothem is the long leg thus $a = 3\sqrt{3}$.

$$A_{hexagon} = \frac{1}{2}\left(3\sqrt{3}\right)(36)\,54\sqrt{3}$$

Notice the measurements of the flower bed are in both feed and inches. We will convert 2 inches to part of a foot.

$$2 \text{ inches} = \frac{2}{12} = \frac{1}{6} \text{ ft}$$

$$V = \left(54\sqrt{3}\right)\left(\frac{1}{6}\right) = 9\sqrt{3}$$

Thus, the amount of dirt needed, rounded to the nearest tenth, is approximately 15.6 ft^3.

16. The lateral area of this pyramid is given by $LA = \frac{1}{2}p\ell$ where $p = 3(251) = 753$ cm is the perimeter of its base and $\ell = 314$ cm is its slant height. Thus,

$$LA = \frac{1}{2}p\ell = \frac{1}{2}(753)(314) = 118,221 \text{ cm}^2.$$

17. The formula for the lateral area of a pyramid is $LA = \frac{1}{2}p\ell$. Since the figure is a regular hexagon, the perimeter is $p = 6(9.8) = 58.8$ ft.

The apothem is found by using the 30°-60°-90° Triangle Theorem.

The apothem is the long leg, thus $a = 4.9\sqrt{3}$.

Use the Pythagorean Theorem to find the slant height.

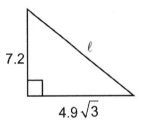

$$\left(7.2\right)^2 + \left(4.9\sqrt{3}\right)^2 = \ell^2$$

$$51.84 + 72.03 = \ell^2$$

$$123.87 = \ell^2$$

$$\ell = \sqrt{123.87}$$

$$LA = \frac{1}{2}(58.8)\left(\sqrt{123.87}\right)$$

The lateral area, rounded to the nearest tenth, is approximately 327.2 ft^2.

18. The formula for the volume of a regular pyramid is $V = \frac{1}{3}Bh$.

The base is an equilateral triangle with side measuring 16.8 in.

The formula for the area of an equilateral triangle is

$$A = \frac{s^2\sqrt{3}}{4}.$$

The area of the base is

$$A = \frac{(16.8)^2\sqrt{3}}{4} = 70.56\sqrt{3}.$$

$$V = \frac{1}{3}\left(70.56\sqrt{3}\right)(25.5)$$

Thus, the volume of the pyramid, to the nearest tenth, is approximately 1038.8 in^3.

19. The formula for the volume of a regular pyramid is $V = \frac{1}{3}Bh$. Since the base is a square, the area is $A = s^2 = (42.6)^2 = 1814.76$. Use the Pythagorean Theorem to find the height of the pyramid.

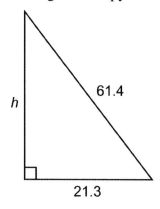

$$h^2 + (21.3)^2 = (61.4)^2$$
$$h^2 + 453.69 = 3769.96$$
$$h^2 = 3316.27$$
$$h = \sqrt{3316.27}$$
$$V = \frac{1}{3}(1814.76)\left(\sqrt{3316.27}\right)$$

The volume, rounded to the nearest tenth, is approximately 34,835.6 m^3.

20. The formula for the volume of a regular pyramid is needed. $V = \frac{1}{3}Bh$.

Since the base is an equilateral triangle the formula for the area is $A = \frac{a^2\sqrt{3}}{4}$.

$$A = \frac{(2.2)^2\sqrt{3}}{4} = 1.21\sqrt{3}$$

$$V = \frac{1}{3}\left(1.21\sqrt{3}\right)(4.1)$$

Thus the bottle will hold, rounded to the nearest tenth, approximately 2.9 in^3 of liquid.

21. The formula for the surface area of a cylinder is $SA = LA + 2B$ or $2\pi rh + 2\pi r^2$.

$$SA = 2\pi(2.3)(6.5) + 2\pi(2.3)^2$$
$$= 29.9\pi + 10.58\pi$$
$$= 40.48\pi$$

Thus, the surface area of the cylinder, rounded to the nearest tenth is approximately 127.2 ft^2.

22. The formula for the surface area of a cylinder is $SA = LA + 2B$ or $2\pi rh + 2\pi r^2$.

$$SA = 2\pi(21.2)(19.5) + 2\pi(21.2)^2$$
$$= 826.8\pi + 898.88\pi$$
$$= 1725.68\pi$$

Thus, the surface area of the cylinder, rounded to the nearest tenth, is approximately 5421.4 cm^2.

23. The formula for the volume of a cylinder is $V = Bh$ or $\pi r^2 h$.

$$V = \pi(165)^2(214) = 5,826,150\pi$$

Thus, the volume of the cylinder, rounded to the nearest tenth, is approximately 18,303,390.0 in^3.

24. The formula for the volume of a cylinder is $V = Bh$ or $\pi r^2 h$.

$$V = \pi(42.6)^2(15.7) = 28{,}491.732\pi$$

Thus, the volume of the cylinder, rounded to the nearest tenth, is approximately 89,509.4 m^3.

25. The formula for the surface area of a cone is $SA = LA + B$ or $\pi r \ell + \pi r^2$.

$$SA = \pi(6.2)(9.5) + \pi(6.2)^2$$
$$= 58.9\pi + 38.44\pi$$
$$= 97.34\pi$$

Thus, the surface area of the cone, rounded to the nearest tenth, is approximately 305.8 cm^2.

26. The formula for the surface area of a cone is $SA = LA + B$ or $\pi r \ell + \pi r^2$. Use the Pythagorean Theorem to find the slant height (ℓ).

$$(18.9)^2 + (26.8)^2 = \ell^2$$
$$357.21 + 718.24 = \ell^2$$
$$1075.45 = \ell^2$$
$$\ell = \sqrt{1075.45}$$
$$SA = \pi(18.9)\left(\sqrt{1075.45}\right) + \pi(18.9)^2$$

Thus, the surface area of the cone, rounded to the nearest tenth, is approximately 3069.4 ft^2.

27. The formula for the volume of a cone is $V = \frac{1}{3}Bh$ or $\frac{1}{3}\pi r^2 h$.

$$V = \frac{1}{3}\pi(12.8)^2(19.2) = 1048.576\pi$$

Thus, the volume of the cone, rounded to the nearest tenth, is approximately 3294.2 m^3.

28. The formula for the volume of a cone is $V = \frac{1}{3}Bh$ or $\frac{1}{3}\pi r^2 h$. Use the Pythagorean Theorem to find the height (h) of the cone.

$$(25.8)^2 + h^2 = (39.2)^2$$
$$665.64 + h^2 = 1536.64$$
$$h^2 = 871$$
$$h = \sqrt{871}$$
$$V = \frac{1}{3}\pi(25.8)^2\left(\sqrt{871}\right)$$

Thus, the volume of the cone, rounded to the nearest tenth, is approximately 20,572.0 in^3.

29. The volume of the cone is needed. The formula is $V = \frac{1}{3}Bh$ or $\frac{1}{3}\pi r^2 h$.

Since the diameter is 1 foot = 12 inches, the radius is 6 inches.

$$V = \frac{1}{3}\pi(6)^2(10) = 120\pi$$

Thus, the funnel can hold, rounded to the nearest tenth, approximately 377.0 in^3 of oil.

30. The metal tube is 2 meters long, therefore, since 1 m = 100 cm, the tube is 200 cm long. To find the amount of metal used, find the volume of the "outer" cylinder and subtract the volume of the "inner" cylinder. The radius of the "outer" cylinder is 15 cm and the radius of the "inner" cylinder is 10 cm. The formula for the volume of a cylinder is $V = Bh$ or $\pi r^2 h$.

$$V_{outer} = \pi(15)^2(200) = 45{,}000\pi$$
$$V_{inner} = \pi(10)^2(200) = 20{,}000\pi$$
$$45{,}000\pi - 20{,}000\pi = 25{,}000\pi$$

The number of cubic centimeters of metal used to make the tube, rounded to the nearest tenth, is approximately 78,539.8 cm^3.

31. The formula for the surface area of a sphere is $SA = 4\pi r^2$.

$$SA = 4\pi(3.9)^2 = 60.84\pi$$

Thus, the surface area of the sphere, rounded to the nearest tenth, is approximately 191.1 ft^2.

32. The formula for the surface area of a sphere is $SA = 4\pi r^2$.

$$SA = 4\pi(26.8)^2 = 2872.96\pi$$

Thus, the surface area of the sphere, rounded to the nearest tenth, is approximately 9025.7 m^2.

33. The formula for the volume of a sphere is $V = \frac{4}{3}\pi r^3$.

$$V = \frac{4}{3}\pi(0.92)^3$$

Thus, the volume of this sphere, rounded to the nearest tenth, is approximately 3.3 cm^3.

34. The formula for the volume of a sphere is $V = \frac{4}{3}\pi r^3$.

$$V = \frac{4}{3}\pi(12.9)^3$$

Thus, the volume of this sphere, rounded to the nearest tenth, is approximately 8992.0 in^3.

35. The surface area of the ball is needed. The formula for the surface area of a sphere is $SA = 4\pi r^2$. The formula for the circumference of a circle is $C = 2\pi r$. The circumference of the ball is 30 inches, therefore to find the radius of the sphere use the equation: $2\pi r = 30$.

$$r = \frac{15}{\pi}$$

$$SA = 4\pi\left(\frac{15}{\pi}\right)^2$$

Thus, the surface area of the ball, rounded to the nearest tenth, is approximately 286.5 in^2.

36. To find the radius of the sphere, solve the equation: $4\pi r^2 = 100$.

$$r^2 = \frac{25}{\pi}$$

$$r = \sqrt{\frac{25}{\pi}}$$

Now use the volume formula for a sphere.

$$V = \frac{4}{3}\pi r^3$$

$$= \frac{4}{3}\pi\left(\sqrt{\frac{25}{\pi}}\right)^3$$

Thus, the volume of the sphere, rounded to the nearest tenth, is approximately 94.0 m^2.

37. First find the volume of the spherical tank by using the formula.

$$V = \frac{4}{3}\pi r^3$$

$$= \frac{4}{3}\pi(15.8)^3$$

To find the number of grams of liquid the tank will hold, multiply the volume by 0.9.

$$\text{amount of liquid} = \left(\frac{4}{3}\pi(15.8)^3\right)(0.9)$$

Thus the number of grams of liquid, rounded to the nearest tenth, is approximately 14,869.7 g.

38. Find the volume of the cylindrical part of the bottle and add it to the volume of the conical top. The formula for the volume of a cylinder is $V = Bh$ or $\pi r^2 h$.
Since the diameter of the cylindrical part is 4 cm, the radius is 2 cm.

$$V_{cylinder} = \pi(2)^2(7) = 28\pi$$

The formula for the volume of a cone is $V = \frac{1}{3}Bh$ or $\frac{1}{3}\pi r^2 h$.

$$V_{cone} = \frac{1}{3}\pi(2)^2(3) = 4\pi$$

The total volume is $28\pi + 4\pi = 32\pi$.

Thus, the glue bottle will hold, rounded to the nearest tenth, approximately 100.5 cm^3 of glue.

39. Find the volume of the cylindrical part and add it to the volume of the two hemispheres or one sphere with radius of 1.8 m. Since the overall height of the tank is 9.9, to find the height of the cylinder subtract twice the radius from 9.9:

$$9.9 - 2(1.8) = 6.3 = h.$$

The formula for the volume of a cylinder is $V = Bh$ or $\pi r^2 h$.

$$V_{cylinder} = \pi(1.8)^2(6.3) = 20.412\pi$$

$$V_{sphere} = \frac{4}{3}\pi(1.8)^3 = 7.776\pi$$

Thus, the tank will hold $20.41\pi + 7.776\pi = 28.188\pi$ m^3.

Since one cubic meter of water weighs 1000 kg, multiply the volume by 1000 kg. To the nearest tenth, the number of kilograms of water in the tank is $(28.188\pi)(1000) \approx 88{,}555.2$ kg.

40. The surface area of the machine part is needed which consists of adding the lateral area of the cone to the surface area of the hemisphere. The formula for the lateral area of a cone is $LA = \frac{1}{2}p\ell$ or $\pi r\ell$.

$$LA_{cone} = \pi(2.3)(12.2) = 28.06\pi$$

To find the surface area of a hemisphere, multiply the surface area of a sphere by one-half.

$$SA_{hemisphere} = \frac{1}{2}\left(4\pi r^2\right)$$

$$SA_{hemisphere} = \frac{1}{2}(4\pi)(2.3)^2$$

$$= 10.58\pi$$

The surface area of the machine part is $28.06\pi + 10.58\pi = 38.64\pi$ cm^2.

To find the cost to finish the part, multiply the surface area by \$2.65.

The cost to finish the part, rounded to the nearest cent, is approximately $(38.64\pi)(2.65) \approx \$321.69$.

Chapter 8 Practice Test

1. True   **2.** True

3. True   **4.** True

5. The surface area of the tank is needed. Use the formula for the surface area of a sphere.

$$SA = 4\pi r^2$$

$$SA = 4\pi(8.6)^2 = 295.84\pi \text{ m}^2$$

To find the cost of the rustproofing, multiply the surface area times $1.50.

The cost to rustproof the tank, rounded to the nearest cent, is approximately $(295.84\pi)(1.50) \approx \$1394.11$.

6. The formula for the surface area of a right prism is $SA = LA + 2B$ where $LA = ph$.

The perimeter of the equilateral triangle is $p = 3(12.6) = 37.8$.

The formula for the area of an equilateral triangle is $A = \dfrac{a^2\sqrt{3}}{4}$.

The area of one base is $\dfrac{(12.6)^2\sqrt{3}}{4} = 39.69\sqrt{3}$.

$$SA_{prism} = 37.8(8.2) + 2\left(39.69\sqrt{3}\right)$$

Thus, the surface area of the prism, rounded to the nearest tenth, is approximately 447.5 cm^2.

7. The formula for the volume of a right prism is $V = Bh$.

Since the base is a rectangle, its area is $A = bh = (4.1)(9.2) = 37.72$.

$$V_{prism} = (37.72)(5.5) = 207.46$$

Thus, the exact volume of the prism is 207.46 m^3.

8. The formula for the volume of a cone is $V = \frac{1}{3}Bh$ or $\frac{1}{3}\pi r^2 h$.

Use the Pythagorean Theorem to find the height (h) of the cone.

$$(28.6)^2 + h^2 = (42.5)^2$$
$$817.96 + h^2 = 1806.25$$
$$h^2 = 988.29$$
$$h = \sqrt{988.29}$$
$$V_{cone} = \frac{1}{3}\pi(28.6)^2\left(\sqrt{988.29}\right)$$

Thus the volume of the cone, rounded to the nearest tenth, is approximately $26{,}927.9$ in^3.

9. The formula for the lateral area of a regular pyramid is $LA = \frac{1}{2}p\ell$.

The perimeter of the square base is $p = 4(32.2) = 128.8$.

Use the Pythagorean Theorem to find the slant height (ℓ).

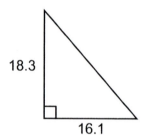

One leg of the right triangle is one-half the side of the square base.

$$(18.3)^2 + (16.1)^2 = \ell^2$$
$$334.89 + 259.21 = \ell^2$$
$$594.1 = \ell^2$$
$$\ell = \sqrt{594.1}$$
$$LA_{pyramid} = \frac{1}{2}(128.8)\left(\sqrt{594.1}\right)$$

Thus, the lateral area of the pyramid, rounded to the nearest tenth, is approximately 1569.7 m^2.

10. The formula for the volume of a pyramid is $V = \frac{1}{3}Bh$.

The base is a square with area $A = (32.2)^2 = 1036.84$.

$$V_{pyramid} = \frac{1}{3}(1036.84)(18.3) = 6324.724$$

Thus, the exact volume of the pyramid is 6324.724 m^3.

11. The formula for the surface area of a cylinder is $SA = LA + 2B$ or $2\pi rh + 2\pi r^2$.

$$SA = 2\pi(1.2)(6.7) + 2\pi(1.2)^2$$
$$= 16.08\pi + 2.88\pi = 18.96\pi$$

Thus, the surface area of the cylinder, rounded to the nearest tenth, is approximately 59.6 ft^2.

12. The formula for the volume of a cylinder is $V = Bh$ or $\pi r^2 h$.

$$V_{cylinder} = \pi(1.2)^2(6.7) = 9.648\pi$$

Thus, the volume of the cylinder, rounded to the nearest tenth, is approximately 30.3 ft^3.

13. The formula for the surface area of a sphere is $SA = 4\pi r^2$.

Since the diameter is 14 cm, the radius is 7 cm.

$$SA = 4\pi(7)^2 = 196\pi$$

Thus, the surface area of the sphere, rounded to the nearest tenth, is approximately 615.8 cm^2.

14. The formula for the volume of a prism is $V = Bh$.

The formula for the area of a hexagon is $A = \frac{1}{2}ap$.

The perimeter of the regular hexagon is $p = 6(8) = 48$.

Use the 30°-60°-90° Triangle Theorem to find the apothem.

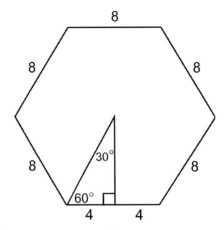

The apothem is the long leg in the right triangle thus $a = 4\sqrt{3}$.

$$A_{hexagon} = \frac{1}{2}(4\sqrt{3})(48) = 96\sqrt{3}$$
$$V_{prism} = (96\sqrt{3})(12) = 1152\sqrt{3}$$

Thus, the volume of the hexagonal prism, rounded to the nearest tenth, is approximately 1995.3 cm^3.

15. The lateral area of the cone is needed. The formula for the lateral area of a cone is $LA = \frac{1}{2}p\ell$ or $\pi r\ell$.

Since the diameter of the cone is 14 inches, the radius is 7 inches.

Use the Pythagorean Theorem to find the slant height (ℓ) of the cone.

$$7^2 + 24^2 = \ell^2$$
$$49 + 576 = \ell^2$$
$$625 = \ell^2$$
$$\ell = 25$$
$$LA_{cone} = \pi(7)(25) = 175\pi$$

Thus, rounded to the nearest tenth, 549.8 in^2 of plastic is needed to make the ice cream cone.

16. The volume of the cylindrical part of the silo must be added to the volume of the hemisphere. To find the height of the cylindrical part of the silo, subtract the radius of the hemisphere from the overall height: $57.2 - 9.8 = 47.4$ m.

The formula for the volume of a cylinder is $V = Bh$ or $\pi r^2 h$.

$$V_{cylinder} = \pi (9.8)^2 (47.4) = 4552.296\pi$$

The formula for the volume of a hemisphere is one-half the volume of a sphere: $V = \frac{1}{2} \cdot \frac{4}{3}\pi r^3$.

$$V_{hemisphere} = \frac{1}{2} \cdot \frac{4}{3}\pi (9.8)^3$$

$$V_{silo} = 4552.296\pi + \frac{1}{2} \cdot \frac{4}{3}\pi (9.8)^3$$

Thus, the amount of grain the silo will hold, rounded to the nearest tenth, is approximately 16,272.7 m³.

17. The formula for the surface area of a cone is $SA = LA + B$ or $\pi r \ell + \pi r^2$.

Use the Pythagorean Theorem to find the slant height (ℓ).

$$3^2 + 4^2 = \ell^2$$

$$9 + 16 = \ell^2$$

$$25 = \ell^2$$

$$\ell = 5$$

$$SA_{cone} = \pi(3)(5) + \pi(3)^2$$

$$= 15\pi + 9\pi = 24\pi$$

Thus, the surface area of the cone, rounded to the nearest tenth, is approximately 75.4 in².

18. To calculate the amount of air in the ball, volume is needed. The formula for the volume is $V = \frac{4}{3}\pi r^3$. Since the circumference of the ball is 27 inches, to find the radius solve:

$$2\pi r = 27$$

$$r = \frac{27}{2\pi}$$

$$V_{sphere} = \frac{4}{3}\pi \left(\frac{27}{2\pi}\right)^3.$$

Thus, rounded to the nearest tenth, the amount of air in the ball is approximately 332.4 in³.

CHAPTER 9 ANALYTIC GEOMETRY

Section 9.1 The Cartesian Coordinate System, Distance and Midpoint Formulas

9.1 PRACTICE EXERCISES

1. Graph $2x - 3y = 6$.

Complete the intercept table:

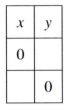

x	y
0	
	0

When $x = 0$.

$$2(0) - 3y = 6$$
$$-3y = 6$$
$$y = -2.$$

When $y = 0$.

$$2x - 3(0) = 6$$
$$2x = 6$$
$$x = 3.$$

Thus, the completed table is

x	y
0	−2
3	0

The graph is the straight line through the points $(0, -2)$ and $(3, 0)$ shown below.

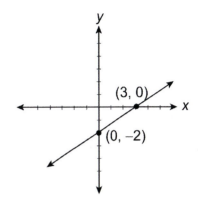

2. Use the distance formula:

$$\sqrt{(x_2 - x_1)^2 + (y_2 - y_1)^2}$$

Let $x_2 = -3, x_1 = 2$ and $y_2 = 5, y_1 = 3$;

$$d = \sqrt{(-3-2)^2 + (5-3)^2}$$
$$d = \sqrt{(-5)^2 + (2)^2}$$
$$= \sqrt{25+4}$$
$$= \sqrt{29}$$

Use the midpoint formula.

$$\left(\frac{x_1 + x_2}{2}, \frac{y_1 + y_2}{2}\right) = \left(\frac{2+(-3)}{2}, \frac{3+5}{2}\right)$$
$$= \left(-\frac{1}{2}, 4\right)$$

9.1 SECTION EXERCISES

1.

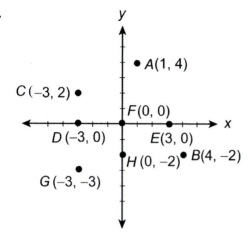

3. $A(3,5), B(4,1), C(0,0), D(-3,2), E(-5,0),$
$F(0,-4), G(-4,-6), H(5,-4)$

5.

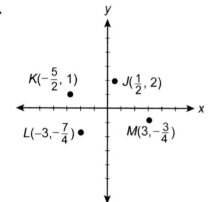

7. J: I, K: II, L: III, M: IV

9. (a) Substitute 0 for x and solve for y.

$$x + y + 2 = 0$$
$$0 + y + 2 = 0$$
$$y = -2$$

(b) Substitute 0 for y and solve for x.

$$x + y + 2 = 0$$
$$x + 0 + 2 = 0$$
$$x = -2$$

(c) Substitute 1 for x and solve for y.

$$x + y + 2 = 0$$
$$1 + y + 2 = 0$$
$$y + 3 = 0$$
$$y = -3$$

(d) Substitute -2 for y and solve for x.

$$x + y + 2 = 0$$
$$x + (-2) + 2 = 0$$
$$x + 0 = 0$$
$$x = 0$$

11. (a) Since $x + 3 = 0, x = -3$. Thus, it is impossible for x to be 0.

(b) Since $x + 3 = 0, x = -3$. Thus, regardless of the value for y, including $y = 0, x = -3$.

(c) Since $x + 3 = 0, x = -3$. Thus, it is impossible for x to be 2.

(d) Since $x + 3 = 0, x = -3$. Thus, regardless of the value for y, including $y = -4, x = -3$.

13. (a) Since $3y + 1 = 0, 3y = -1$ so that $y = -\dfrac{1}{3}$. Thus, for every value of x, including $x = 0, y = -\dfrac{1}{3}$.

(b) Since $y = -\dfrac{1}{3}$, it is impossible for y to be 0.

(c) Since $y = -\dfrac{1}{3}$ for all values of x, including $x = 3, y = -\dfrac{1}{3}$.

(d) Since $y = -\dfrac{1}{3}$, it is impossible for y to be -1.

15. x-intercept $(-2,0)$;
  y-intercept $(0,-2)$

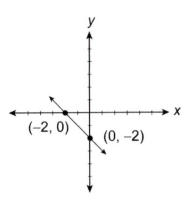

17. x-intercept $(2,0)$;
  y-intercept $(0,6)$

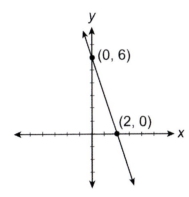

19. x-intercept $(2,0)$;
  y-intercept $(0,-2)$

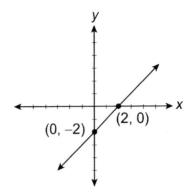

21. x-intercept $(0,0)$;
  y-intercept $(0,0)$

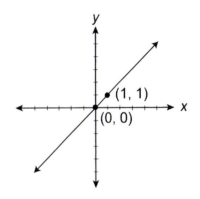

23. x-intercept $(\dfrac{7}{3},0)$;
  no y-intercept

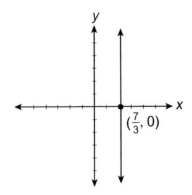

25. no x-intercept
y-intercept $(0,-1)$;

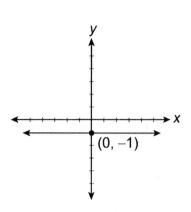

27. x-intercept $(0,0)$;
y-intercept $(0,0)$;

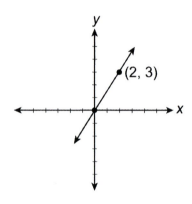

29. x-intercept $(2,0)$;
y-intercept $(0,3)$

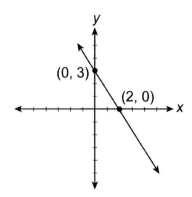

31. x-intercept $(0,0)$; every point on the y-axis is
a y-intercept; the graph is the y-axis.

33. All have the same y-intercept, $(0,1)$.

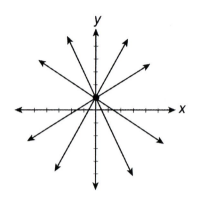

35. $d = \sqrt{(x_2-x_1)^2+(y_2-y_1)^2}$
$= \sqrt{(-2-3)^2+(-10-2)^2}$
$= \sqrt{(-5)^2+(-12)^2}$
$= \sqrt{25+144} = \sqrt{169} = 13$

37. $d = \sqrt{(x_2-x_1)^2+(y_2-y_1)^2}$
$= \sqrt{(-3-0)^2+(0-2)^2}$
$= \sqrt{(-3)^2+(-2)^2}$
$= \sqrt{9+4} = \sqrt{13}$

39. $d = \sqrt{(x_2-x_1)^2+(y_2-y_1)^2}$
$= \sqrt{(5-5)^2+(-3-2)^2}$
$= \sqrt{(0)^2+(-5)^2}$
$= \sqrt{0+25} = \sqrt{25} = 5$

41. $d = \sqrt{(x_2-x_1)^2+(y_2-y_1)^2}$
$10 = \sqrt{(4-4)^2+(f-3)^2}$
$10 = \sqrt{0+(f-3)^2}$
$10 = f-3$
$13 = f$

43. $\text{midpoint} = \left(\dfrac{x_1 + x_2}{2}, \dfrac{y_1 + y_2}{2} \right)$

$\left(\dfrac{-2+4}{2}, \dfrac{-1+1}{2} \right) = \left(\dfrac{2}{2}, \dfrac{0}{2} \right) = (1, 0)$

45. $\text{midpoint} = \left(\dfrac{x_1 + x_2}{2}, \dfrac{y_1 + y_2}{2} \right)$

$\left(\dfrac{6+5}{2}, \dfrac{11+(-3)}{2} \right) = \left(\dfrac{11}{2}, \dfrac{8}{2} \right) = (\dfrac{11}{2}, 4)$

47. $\text{midpoint} = \left(\dfrac{x_1 + x_2}{2}, \dfrac{y_1 + y_2}{2} \right)$

$\left(\dfrac{2k+2p}{2}, \dfrac{2n+2q}{2} \right) = (k+p, \ n+q)$

49. $\text{midpoint} = \left(\dfrac{x_1 + x_2}{2}, \dfrac{y_1 + y_2}{2} \right)$

$(0,0) = \left(\dfrac{2+x_2}{2}, \dfrac{-4+y_2}{2} \right)$

$\dfrac{2+x_2}{2} = 0 \qquad \dfrac{-4+y_2}{2} = 0$

$x_2 = 0 \qquad\qquad y_2 = 4$

Thus the other endpoint is $(-2, 4)$.

51. (a) $\text{midpoint} = \left(\dfrac{x_1 + x_2}{2}, \dfrac{y_1 + y_2}{2} \right)$

$= \left(\dfrac{0+6}{2}, \dfrac{8+0}{2} \right)$

$= (3, 4)$

(b) $d = \sqrt{(x_2 - x_1)^2 + (y_2 - y_1)^2}$

$AM = \sqrt{(3-0)^2 + (4-8)^2}$

$= \sqrt{3^2 + (-4)^2}$

$= \sqrt{9+16}$

$= \sqrt{25} = 5$

$MC = \sqrt{(6-3)^2 + (0-4)^2}$

$= \sqrt{3^2 + (-4)^2}$

$= \sqrt{9+16}$

$= \sqrt{25} = 5$

$MB = \sqrt{(3-0)^2 + (4-0)^2}$

$= \sqrt{3^2 + 4^2}$

$= \sqrt{9+16}$

$= \sqrt{25} = 5$

Theorem 5.16 states the median ($\overline{MB}$) is one-half the length of the hypotenuse ($\overline{AC}$). These calculations support the Theorem:

$\dfrac{AC}{2} = \dfrac{5+5}{2} = 5 = BM.$

53. (a) $d = \sqrt{(x_2 - x_1)^2 + (y_2 - y_1)^2}$

$AB = \sqrt{(5-1)^2 + (2-2)^2} = \sqrt{4^2} = 4$

$BC = \sqrt{(1-5)^2 + (2-6)^2}$

$= \sqrt{(-4)^2 + (-4)^2}$

$= \sqrt{16+16} = \sqrt{32} \text{ or } 4\sqrt{2}$

$AC = \sqrt{(5-5)^2 + (6-2)^2} = \sqrt{4^2} = 4$

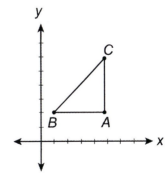

(b) It is a right isosceles $\triangle$ because 2 sides are congruent and $(AB)^2 + (AC)^2 = (BC)^2$.

55. Answers will vary.

Section 9.2 Slope, Equation of a Line

9.2 PRACTICE EXERCISES

1. Use the formula for slope

$$m = \frac{y_2 - y_1}{x_2 - x_1}$$

with $(x_1, y_1) = (1, -3)$ and $(x_2, y_2) = (-2, 4)$.

$$m = \frac{4 - (-3)}{-2 - 1} = \frac{4 + 3}{-3} = \frac{7}{-3} = -\frac{7}{3}$$

2. Use the point-slope form $y - y_1 = m(x - x_1)$
with $m = -3$ and $(x_1, y_1) = (2, -3)$

$$y - (-3) = -3(x - 2)$$
$$y + 3 = -3(x - 2)$$
$$y + 3 = -3x + 6$$
$$3x + y - 3 = 0$$

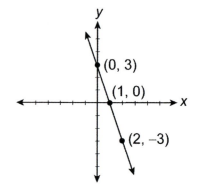

9.2 SECTION EXERCISES

1. $m = \dfrac{y_2 - y_1}{x_2 - x_1} = \dfrac{-6 - 2}{-1 - (-5)} = \dfrac{-8}{4} = -2$

3. $m = \dfrac{y_2 - y_1}{x_2 - x_1} = \dfrac{7 - 7}{-5 - (-2)} = \dfrac{0}{-3} = 0$

5. $m = \dfrac{y_2 - y_1}{x_2 - x_1} = \dfrac{b - 0}{0 - a} = \dfrac{b}{-a}$ or $-\dfrac{b}{a}$

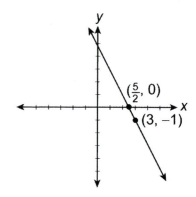

7. Use the point-slope form $y - y_1 = m(x - x_1)$
with the point $(3, -1)$ and the slope -2.

$$y - y_1 = m(x - x_1)$$
$$y - (-1) = -2(x - 3)$$
$$y + 1 = -2x + 6$$
$$2x + y - 5 = 0$$

9. A line parallel to a line with slope $\frac{1}{3}$ also has
slope $\frac{1}{3}$. Use the point-slope form.

$$y - y_1 = m(x - x_1)$$
$$y - (-4) = \frac{1}{3}(x - 2)$$
$$3(y + 4) = 1(x - 2)$$
$$3y + 12 = x - 2$$
$$0 = x - 3y - 14$$
$$x - 3y - 14 = 0$$

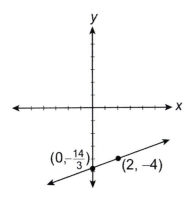

11. First find the slope of the line.

$$m = \frac{y_2 - y_1}{x_2 - x_1} = \frac{3 - (-1)}{5 - 7} = \frac{3 + 1}{-2} = \frac{4}{-2} = -2$$

Now use the point-slope form with $(7, -1)$ as the point and -2 as the slope.

$$y - y_1 = m(x - x_1)$$
$$y - (-1) = -2(x - 7)$$
$$y + 1 = -2x + 14$$
$$2x + y - 13 = 0$$

Note: the point $(5, 3)$ could have been used in the point-slope form with the same results.

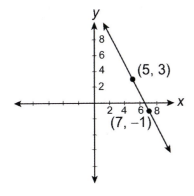

13. A vertical line has undefined slope. Since every point on the line has the same x-coordinate, 2, the desired equation is $x = 2$ or $x - 2 = 0$.

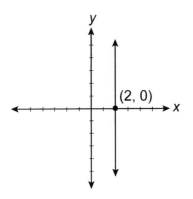

15. A line with slope 0 is a horizontal line. We could use the point-slope form with $(2, 3)$ as the point.

$$y - y_1 = m(x - x_1)$$
$$y - 3 = 0(x - 2)$$
$$y - 3 = 0$$

Another way to think of this graph is that every point on a horizontal line has the same y-coordinate, here it is 3. Thus the desired equation is $y = 3$ or $y - 3 = 0$.

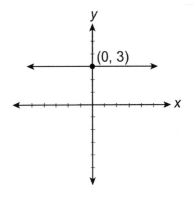

17. Use the point-slope form with $(-5,0)$ as the point and 2 as the slope.

$$y - y_1 = m(x - x_1)$$
$$y - 0 = 2(x - (-5))$$
$$y = 2(x + 5)$$
$$y = 2x + 10$$
$$0 = 2x - y + 10$$
or $2x - y + 10 = 0$

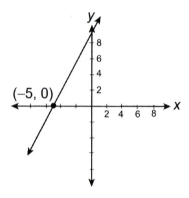

19. Since the y-intercept is $(0,4)$ and the slope is -3, we can use the slope-intercept form.

$$y = mx + b$$
$$y = -3x + 4$$
$$3x + y - 4 = 0$$

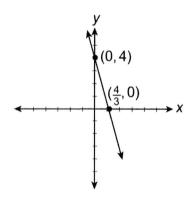

21. First find the slope.

$$m = \frac{y_2 - y_1}{x_2 - x_1} = \frac{\frac{2}{3} - (-1)}{7 - 4} = \frac{\frac{2}{3} + 1}{3} = \frac{\frac{5}{3}}{3}$$
$$= \frac{5}{3} \cdot \frac{1}{3} = \frac{5}{9}$$

Now use the point-slope form with $(4,-1)$ as the point and $\frac{5}{9}$ as the slope.

$$y - y_1 = m(x - x_1)$$
$$y - (-1) = \frac{5}{9}(x - 4)$$
$$y + 1 = \frac{5}{9}(x - 4)$$
$$9(y + 1) = 5(x - 4)$$
$$9y + 9 = 5x - 20$$
$$0 = 5x - 9y - 29$$
or $5x - 9y - 29 = 0$

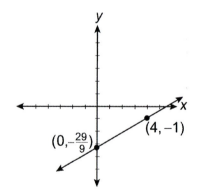

23. First find the slope.

$$m = \frac{y_2 - y_1}{x_2 - x_1} = \frac{5 - 0}{0 - (-2)} = \frac{5}{2}$$

Now use the point-slope form with $(-2,0)$ as the point and $\frac{5}{2}$ as the slope.

$$y - y_1 = m(x - x_1)$$

$$y - 0 = \frac{5}{2}(x - (-2))$$

$$y = \frac{5}{2}(x + 2)$$

$$2y = 5(x + 2)$$

$$2y = 5x + 10$$

$$0 = 5x - 2y + 10$$

or $\quad 5x - 2y + 10 = 0$

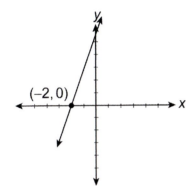

25. Solve the equation for y to write it in slope-intercept form.

$$5x - 2y + 4 = 0$$

$$-2y = -5x - 4$$

$$y = \frac{5}{2}x + 2$$

Then the slope is $\frac{5}{2}$, and the y-intercept is $(0, 2)$.

27. Since there is no y-term in $5x + 4 = 0$, we cannot write the equation solved for y in slope-intercept form. This is a vertical line through x-intercept $\left(-\frac{4}{5}, 0\right)$. It has no y-intercept and the slope is undefined.

29. Solve the equation for y to write it in slope-intercept form.

$$-2y + 4 = 0$$

$$-2y = -4$$

$$y = 2$$

$$y = 0x + 2$$

Thus, the slope is 0, and the y-intercept is $(0, 2)$.

31. First find the slope of the line.

$$m = \frac{y_2 - y_1}{x_2 - x_1} = \frac{-3 - 3}{1 - (-2)} = \frac{-6}{1 + 2} = \frac{-6}{3} = -2$$

Now use the point-slope form with $(-2, 3)$ as the point and -2 as the slope.

$$y - y_1 = m(x - x_1)$$

$$y - 3 = -2(x - (-2))$$

$$y - 3 = -2(x + 2)$$

$$y - 3 = -2x - 4$$

$$y = -2x - 1$$

Thus, the slope-intercept form is $y = -2x - 1$, the slope is -2, and the y-intercept is $(0, -1)$.

Note: The point $(1, -3)$ could have been used in the point-slope form with the same results.

33. Solve each equation for y writing it in slope-intercept form.

$$5x - 3y + 8 = 0 \qquad\qquad 3x + 5y - 7 = 0$$

$$-3y = -5x - 8 \qquad\qquad 5y = -3x + 7$$

$$y = \frac{5}{3}x + \frac{8}{3} \qquad\qquad y = -\frac{3}{5}x + \frac{7}{5}$$

The slope is $\frac{5}{3}$. $\qquad\qquad$ The slope is $-\frac{3}{5}$.

Since $\left(\frac{5}{3}\right)\left(-\frac{3}{5}\right) = -1$, the lines are perpendicular.

35. Solve each equation for y writing it in slope-intercept form.

$$2x - y + 7 = 0 \qquad -6x + 3y - 1 = 0$$
$$-y = -2x - 7 \qquad 3y = 6x + 1$$
$$y = 2x + 7 \qquad y = 2x + \frac{1}{3}$$

The slope is 2   The slope is 2.

Since both slopes are 2 and the y-intercepts are different, the lines are parallel.

37. $\quad 2x + 1 = 0 \qquad \quad -3y - 4 = 0$
$$2x = -1 \qquad \qquad -3y = 4$$
$$x = -\frac{1}{2} \qquad \qquad y = -\frac{4}{3}$$

A vertical line through $\left(-\frac{1}{2}, 0\right)$. A

horizontal line through $\left(0, -\frac{4}{3}\right)$. Thus, the

lines are perpendicular.

39. A line parallel to the line with equation $-3x - y + 4 = 0$ has the same slope as this line. First find that slope by writing the equation in slope-intercept form.

$$-3x - y + 4 = 0$$
$$-y = 3x - 4$$
$$y = -3x + 4$$

Since the slope is -3, use the point-slope form with the point $(-1, 4)$.

$$y - y_1 = m(x - x_1)$$
$$y - 4 = -3(x - (-1))$$
$$y - 4 = -3(x + 1)$$
$$y - 4 = -3x - 3$$
$$3x + y - 1 = 0$$

41. A line perpendicular to the line with equation $-3x - y + 4 = 0$ has slope the negative reciprocal of the slope of this line. First find the slope of the given line by writing the equation in slope-intercept form.

$$-3x - y + 4 = 0$$
$$-y = 3x - 4$$
$$y = -3x + 4$$

The slope of this line is -3 so the slope of the desired line is $\frac{1}{3}$. Use the point-slope form with the point $(-1, 4)$.

$$y - y_1 = m(x - x_1)$$
$$y - 4 = \frac{1}{3}(x - (-1))$$
$$y - 4 = \frac{1}{3}(x + 1)$$
$$3(y - 4) = x + 1$$
$$3y - 12 = x + 1$$
$$0 = x - 3y + 13$$
$$\text{or} \quad x - 3y + 13 = 0$$

43. First find the slope of the line through the given points.

$$m = \frac{y_2 - y_1}{x_2 - x_1} = \frac{-6 - 2}{3 - (-3)} = \frac{-8}{6} = -\frac{4}{3}$$

Then the slope of a line perpendicular to this line is the negative reciprocal of $-\frac{4}{3}, \frac{3}{4}$. Now find the midpoint of the segment joining the given points.

$$\text{midpoint} = \left(\frac{x_1 + x_2}{2}, \frac{y_1 + y_2}{2} \right)$$

$$= \left(\frac{-3 + 3}{2}, \frac{2 + (-6)}{2} \right)$$

$$= \left(\frac{0}{2}, \frac{-4}{2} \right)$$

$$= (0, -2)$$

Use the point-slope form with $(0, -2)$ as the point and $\frac{3}{4}$ as the slope.

$$y - y_1 = m(x - x_1)$$

$$y - (-2) = \frac{3}{4}(x - 0)$$

$$y + 2 = \frac{3}{4}x$$

$$4(y + 2) = 3x$$

$$4y + 8 = 3x$$

$$0 = 3x - 4y - 8$$

or $\quad 3x - 4y - 8 = 0$

45. First find the slope of the line through the given points.

$$m = \frac{y_2 - y_1}{x_2 - x_1} = \frac{1 - (-3)}{7 - 5} = \frac{1 + 3}{2} = \frac{4}{2} = 2$$

Then the slope of the line perpendicular to this line is the negative reciprocal of 2, $-\frac{1}{2}$.

Now find the midpoint of the segment joining the given points.

$$\text{midpoint} = \left(\frac{x_1 + x_2}{2}, \frac{y_1 + y_2}{2} \right)$$

$$= \left(\frac{5 + 7}{2}, \frac{-3 + 1}{2} \right)$$

$$= \left(\frac{12}{2}, \frac{-2}{2} \right)$$

$$= (6, -1)$$

Use the point-slope form with $(6, -1)$ as the point and $-\frac{1}{2}$ as the slope.

$$y - y_1 = m(x - x_1)$$

$$y - (-1) = -\frac{1}{2}(x - 6)$$

$$y + 1 = -\frac{1}{2}(x - 6)$$

$$2(y + 1) = (-1)(x - 6)$$

$$2y + 2 = -x + 6$$

$$x + 2y - 4 = 0$$

47. (a) By Postulate 6.4 a radius drawn to a point of tangency is perpendicular to the tangent. Using the slope formula

$$m = \frac{y_2 - y_1}{x_2 - x_1}$$

$$\text{slope of the radius} = \frac{5 - 0}{3 - 0} = \frac{5}{3}.$$

Thus the slope of the tangent line is $-\frac{3}{5}$.

(b) Use the point-slope form where $m = -\frac{3}{5}$ and the point is $(3, 5)$.

$$y - y_1 = m(x - x_1)$$

$$y - 5 = -\frac{3}{5}(x - 3)$$

$$y - 5 = -\frac{3}{5}x + \frac{9}{5}$$

$$y = -\frac{3}{5}x + \frac{9}{5} + 5$$

$$y = -\frac{3}{5}x + \frac{34}{5}$$

49. (a) Use midpoint formula $= \left(\dfrac{x_1 + x_2}{2}, \dfrac{y_1 + y_2}{2} \right)$

midpoint of one leg $= \left(\dfrac{-1 + (-4)}{2}, \dfrac{4 + (-3)}{2} \right)$

$\qquad\qquad\qquad = \left(-\dfrac{5}{2}, \dfrac{1}{2} \right)$

midpoint of other leg $= \left(\dfrac{4 + 7}{2}, \dfrac{4 + (-3)}{2} \right)$

$\qquad\qquad\qquad = \left(\dfrac{11}{2}, \dfrac{1}{2} \right)$

(b) Use the slope formula: $m = \dfrac{y_2 - y_1}{x_2 - x_1}$

slope of one base $= \dfrac{4 - 4}{4 - (-1)} = \dfrac{0}{5} = 0$

slope of other base $= \dfrac{-3 - (-3)}{7 - (-4)} = \dfrac{0}{11} = 0$

slope of median $= \dfrac{\frac{1}{2} - \frac{1}{2}}{\frac{11}{2} - \frac{-5}{2}} = \dfrac{0}{8} = 0$

Since the slopes of the two bases and the median are 0, the lines are parallel.

51. To use the form $y - y_1 = m(x - x_1)$ the slope (m) and a point (x_1, y_1) are needed hence the name point-slope.
To use the form $y = mx + b$, the slope (m) and the y-intercept (b) are needed hence the name slope-intercept.

Section 9.3 Proofs Involving Polygons

9.3 SECTION EXERCISES

1. $C(x, -y)$

3. $A(-q, 0); B(q, 0); C(0, p)$

5. $J(0, y); K(x, y); L(w, 0)$

Midpoint of $\overline{JK}$ is $\left(\dfrac{x}{2}, y \right)$;

Midpoint of $\overline{ML}$ is $\left(\dfrac{w}{2}, 0 \right)$.

7. $C(a - x, y); D(a, 0)$

9. The coordinates of C are (b, y) and the coordinates of D are $(c, 0)$.
Use the distance formula,

$d = \sqrt{(x_2 - x_1)^2 + (y_2 - y_1)^2}$ to find BD and AC.

$$BD = \sqrt{(c - a)^2 + (0 - y)^2}$$
$$= \sqrt{(c - a)^2 + y^2}$$

Substitute $a + b = c$

$$= \sqrt{(a + b - a)^2 + y^2}$$
$$= \sqrt{b^2 + y^2}$$

$$AC = \sqrt{(b - 0)^2 + (y - 0)^2}$$
$$= \sqrt{b^2 + y^2}$$

Thus the diagonals of an isosceles trapezoid are congruent.

11. Draw a figure.
Use the distance formula to prove $AC = BD$.

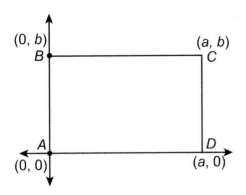

$$AC = \sqrt{(a-0)^2 + (b-0)^2}$$
$$= \sqrt{a^2 + b^2}$$
$$BD = \sqrt{(a-0)^2 + (b-0)^2}$$
$$= \sqrt{a^2 + (-b)^2}$$
$$= \sqrt{a^2 + b^2}$$

Thus the diagonals of a rectangle are $\cong$.

13. Draw an isosceles trapezoid.
Because the figure is isosceles, $a - b = d$.

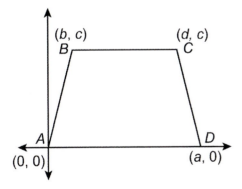

Use the distance formula to show $AC = BD$.

$$AC = \sqrt{(d-0)^2 + (c-0)^2}$$
$$= \sqrt{d^2 + c^2}$$
$$BD = \sqrt{(a-b)^2 + (0-c)^2}$$
$$= \sqrt{(a-b)^2 + (-c)^2}$$
$$= \sqrt{(a-b)^2 + c^2}$$

Substitute $a - b = d$

$$BD = \sqrt{d^2 + c^2}$$

Thus $AC = BD$ and the diagonals of an isosceles trapezoid are $\cong$.

15. Draw a square.

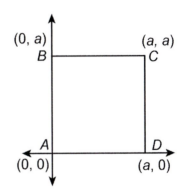

Use the midpoint formula to show the diagonals bisect each other.

$$\text{midpoint of } \overline{AC} = \left(\frac{a+0}{2}, \frac{a+0}{2} \right)$$
$$= \left(\frac{a}{2}, \frac{a}{2} \right)$$

$$\text{midpoint of } \overline{BD} = \left(\frac{0+a}{2}, \frac{a+0}{2} \right)$$
$$= \left(\frac{a}{2}, \frac{a}{2} \right)$$

Thus $\left(\dfrac{a}{2},\dfrac{a}{2}\right)$ is the common midpoint of the diagonals, thus diagonals bisect each other. To show the diagonals are perpendicular, use the slope formula

$$\text{slope } \overline{AC} = \frac{a-0}{a-0} = 1$$

$$\text{slope } \overline{BD} = \frac{a-0}{0-a} = -1$$

The diagonals are $\perp$ because the product of the slopes is -1.

17. Draw a figure.

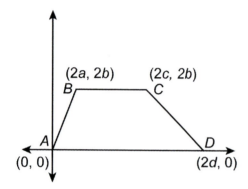

Find the midpoints of $\overline{AB}$ and $\overline{CD}$ using the midpoint formula.

$$\text{midpoint } \overline{AB} = \left(\frac{2a+0}{2},\frac{2b+0}{2}\right)$$
$$= (a,b)$$

Let $M = (a,b)$

$$\text{midpoint } \overline{CD} = \left(\frac{2c+2d}{2},\frac{2b+0}{2}\right)$$
$$= ((c+d),b)$$

Let $N = ((c+d),b)$

The median is $\overline{MN}$. To show the median is parallel to the bases, use the slope formula.

$$\text{slope } \overline{BC} = \frac{2b-2b}{2a-2c} = \frac{0}{2a-2c} = 0$$

$$\text{slope } \overline{MN} = \frac{b-b}{c+d-a} = \frac{0}{c+d-a} = 0$$

$$\text{slope } \overline{AD} = \frac{0-0}{2d-0} = \frac{0}{2d} = 0$$

Since the slopes are equal, the three lines are parallel. The median of a trapezoid is parallel to the bases.

Chapter 9 Review Exercises

1. x-coordinate: -2; y-coordinate: 3; II

2. x-axis and y-axis

3. x-axis

4. y-axis

5. $A(-3,2); B(2,1); C(2,-2); D(-2,-2)$
         II   I   IV   III

6.

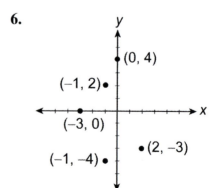

7.

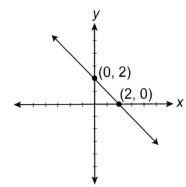

10.

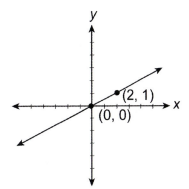

8.

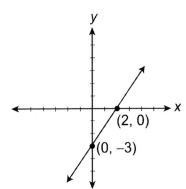

11.

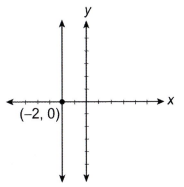

9.

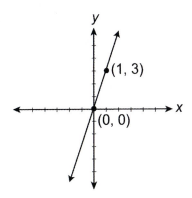

12.

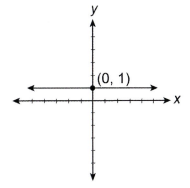

13. $m = \dfrac{y_2 - y_1}{x_2 - x_1} = \dfrac{7 - (-2)}{3 - 1} = \dfrac{7 + 2}{2} = \dfrac{9}{2}$

14. $m = \dfrac{y_2 - y_1}{x_2 - x_1} = \dfrac{-2 - (-4)}{-6 - (-5)} = \dfrac{-2 + 4}{-6 + 5} = \dfrac{2}{-1} = -2$

15. Since parallel lines have equal slopes, a line parallel to a line with slope 5 also has slope 5. A line perpendicular to a line with slope 5 has slope $-\dfrac{1}{5}$ since $\left(-\dfrac{1}{5}\right)(5) = -1.$

16. $m = \dfrac{y_2 - y_1}{x_2 - x_1} = \dfrac{7-(-1)}{-2-3} = \dfrac{7+1}{-5} = -\dfrac{8}{5}$

A line perpendicular to a line with slope $-\dfrac{8}{5}$ has slope $\dfrac{5}{8}$, (the negative reciprocal)

17. $d = \sqrt{(x_2 - x_1)^2 + (y_2 - y_1)^2}$

$= \sqrt{(-2-4)^2 + (-1-(-1))^2}$

$= \sqrt{(-6)^2 + (0)^2} = \sqrt{6^2} = 6$

18. $d = \sqrt{(x_2 - x_1)^2 + (y_2 - y_1)^2}$

$5 = \sqrt{(x-3)^2 + (2-2)^2}$

$5 = \sqrt{(x-3)^2}$

$5 = x-3$

$8 = x$

19. $\text{midpoint} = \left(\dfrac{x_1 + x_2}{2}, \dfrac{y_1 + y_2}{2}\right)$

$= \left(\dfrac{-1+5}{2}, \dfrac{0+(-2)}{2}\right)$

$= \left(\dfrac{4}{2}, \dfrac{-2}{2}\right) = (2, -1)$

20. $\text{midpoint} = \left(\dfrac{x_1 + x_2}{2}, \dfrac{y_1 + y_2}{2}\right)$

$= \left(\dfrac{1+3}{2}, \dfrac{-6+2}{2}\right)$

$= \left(\dfrac{4}{2}, \dfrac{-4}{2}\right) = (2, -2)$

21. To find the perimeter, use the distance formula to find the lengths of the sides.

$AB = \sqrt{(-8-8)^2 + (0-0)^2}$

$ = \sqrt{(-16)^2} = \sqrt{256} = 16$

$AC = \sqrt{(-8-0)^2 + (0-6)^2}$

$ = \sqrt{(-8)^2 + (-6)^2}$

$ = \sqrt{64+36} = \sqrt{100} = 10$

$BC = AC = 10$ since $\triangle ABC$ is isosceles.

$P = 16 + 10 + 10 = 36$

To find the height of the triangle, observe the vertex is $(0,6)$ thus the height is 6, and the base is $\overline{AB}$ where $AB = 16$.

$A = \dfrac{1}{2}bh = \dfrac{1}{2}(16)(6) = 48$

22. Use the point-slope form with $(-1,3)$ as the point and -4 as the slope.

$y - y_1 = m(x - x_1)$

$y - 3 = -4(x-(-1))$

$y - 3 = -4(x+1)$

$y - 3 = -4x - 4$

$4x + y + 1 = 0$

23. Solve the equation for y to find the slope-intercept form.

$4x + 2y - 10 = 0$

$2y = -4x + 10$

$y = -2x + 5$

The slope is -2, and the y-intercept is $(0,5)$.

24. First find the slope of the line.

$$m = \frac{y_2 - y_1}{x_2 - x_1} = \frac{9-5}{6-(-2)} = \frac{4}{6+2} = \frac{4}{8} = \frac{1}{2}$$

Now use the point-slope form with $(-2,5)$ as the point.

$$y - y_1 = m(x - x_1)$$
$$y - 5 = \frac{1}{2}(x - (-2))$$
$$y - 5 = \frac{1}{2}(x + 2)$$
$$2(y - 5) = (1)(x + 2)$$
$$2y - 10 = x + 2$$
$$0 = x - 2y + 12$$
$$\text{so} \quad x - 2y + 12 = 0$$

25. First find the slope of the given line by writing the equation in slope-intercept form.

$$2x + y - 3 = 0$$
$$y = -2x + 3$$

The slope is -2. The slope of a line perpendicular to this line is the negative reciprocal of $-2, \frac{1}{2}$. Use the point-slope form with $(-3,1)$ as the point.

$$y - y_1 = m(x - x_1)$$
$$y - 1 = \frac{1}{2}(x - (-3))$$
$$y - 1 = \frac{1}{2}(x + 3)$$
$$2(y - 1) = (1)(x + 3)$$
$$2y - 2 = x + 3$$
$$0 = x - 2y + 5$$
$$\text{therefore} \quad x - 2y + 5 = 0$$

26. Write the equation in slope-intercept form.

$$3y - 15 = 0$$
$$3y = 15$$
$$y = 5$$
$$y = 0x + 5$$

The slope is 0, and the y-intercept is $(0,5)$. Since the line is horizontal, parallel to the x-axis, there is no x-intercept.

27. Write each equation in slope-intercept form.

$$3x - y + 2 = 0 \qquad\qquad x + 3y - 7 = 0$$
$$-y = -3x - 2 \qquad\qquad 3y = -x + 7$$
$$y = 3x + 2 \qquad\qquad y = -\frac{1}{3}x + \frac{7}{3}$$

The slope is 3.   The slope is $-\frac{1}{3}$.

Since $(3)\left(-\frac{1}{3}\right) = -1$, the lines are perpendicular.

28. First find the slope of the line containing the 2 given points.

$$m = \frac{y_2 - y_1}{x_2 - x_1} = \frac{-4 - (-6)}{8 - (-2)} = \frac{2}{10} = \frac{1}{5}$$

The slope of the line perpendicular to it is -5 (the negative reciprocal). Next find the midpoint between the 2 given points.

$$\text{midpoint} = \left(\frac{x_1 + x_2}{2}, \frac{y_1 + y_2}{2}\right)$$
$$= \left(\frac{-2 + 8}{2}, \frac{-6 + (-4)}{2}\right)$$
$$= (3, -5)$$

Finally, find the equation of the line, in slope-intercept form, with a slope of -5 and contains the point $(3, -5)$.

$$y - y_1 = m(x - x_1)$$
$$y - (-5) = -5(x - 3)$$
$$y + 5 = -5x + 15$$
$$y = -5x + 10$$

29. $(a + b, c)$

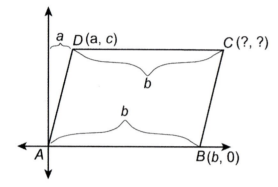

The x-coordinate of C is $a + b$ because the distance from the y-axis to point D is a. The length of $\overline{DC} = b$. Thus the x-coordinate of C is the sum of a and b. The y-coordinate of C will be the same as the y-coordinate in D because $\overline{DC}$ is a horizontal line (parallel to AB).

30. midpoint of $\overline{DC} = \left(\dfrac{a + a + b}{2}, \dfrac{c + c}{2} \right)$

$$= \left(\dfrac{2a + b}{2}, c \right) = M$$

midpoint of $\overline{AB} = \left(\dfrac{0 + b}{2}, \dfrac{0 + 0}{2} \right)$

$$= \left(\dfrac{b}{2}, 0 \right) = N$$

slope $\overline{AD} = \dfrac{c - 0}{a - 0} = \dfrac{c}{a}$

slope $\overline{MN} = \dfrac{c - 0}{\dfrac{2a + b}{2} - \dfrac{b}{2}} = \dfrac{c}{a}$

thus $\overline{AD} \parallel \overline{MN}$ because the slopes are =.

slope $\overline{DM} = \dfrac{c - c}{\dfrac{2a + b}{2} - a} = \dfrac{0}{\dfrac{2a + b}{2} - a} = 0$

slope $\overline{AN} = \dfrac{0 - 0}{\dfrac{b}{2} - 0} = \dfrac{0}{\dfrac{b}{2}} = 0$

Note: there is no need to simplify the denominator since 0 divided by any number (except 0) is 0.

Thus $\overline{DM} \parallel \overline{AN}$ because the slopes are =. Therefore, $ANMD$ is a parallelogram because opposite sides are parallel.

31.

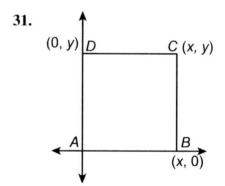

Prove $AC = BD$ using the distance formula.

$$AC = \sqrt{(x - 0)^2 + (y - 0)^2} = \sqrt{x^2 + y^2}$$
$$BD = \sqrt{(x - 0)^2 + (0 - y)^2} = \sqrt{x^2 + y^2}$$

Thus $AC = BD$ therefore $\overline{AC} \cong \overline{BD}$.

32. For convenience, a factor of 2 will be used in the coordinates of B and C.

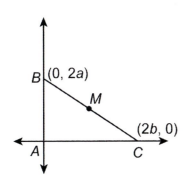

First find the midpoint of $\overline{BC}$.

$$\text{midpoint} = \left(\frac{x_1 + x_2}{2}, \frac{y_1 + y_2}{2} \right)$$
$$= \left(\frac{0 + 2b}{2}, \frac{2a + 0}{2} \right)$$
$$= (b, a)$$

Next show $MB = AM = MC$ using the distance formula.

$$MB = \sqrt{(b-0)^2 + (a-2a)^2} = \sqrt{b^2 + a^2}$$
$$AM = \sqrt{(b-0)^2 + (a-0)^2} = \sqrt{b^2 + a^2}$$
$$MC = \sqrt{(2b-b)^2 + (0-a)^2} = \sqrt{b^2 + a^2}$$

Since the three distances are equal, the midpoint of the hypotenuse is equidistant from the three vertices of the triangle.

Chapter 9 Practice Test

1. Solve the equation for y to write it in slope-intercept form.

$$3x + 2y = 12$$
$$2y = -3x + 12$$
$$y = -\frac{3}{2}x + 6$$

2. To find the x-intercept, substitute 0 for y and solve for x.

$$3x + 2(0) = 12$$
$$3x = 12$$
$$x = 4$$

Thus, the x-intercept is $(4, 0)$.

3. From Problem 1 we see that the y-intercept is $(0, 6)$. This could also be found by substituting 0 for x and solving for y.

4. From Problem 1 we see that the slope is $-\frac{3}{2}$.

5.

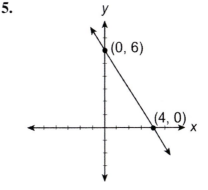

6. We must substitute 10 for x in $y = 4x + 30$.

$$y = 4(10) + 30 = 40 + 30 = 70$$

Then $y = 70$ so the ordered pair solution is $(10, 70)$.

7.

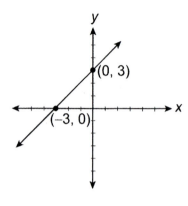

8.

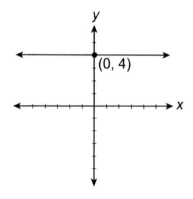

9. x-intercept: $(-5, 0)$;
y-intercept: $(0, -5)$

10. $m = \dfrac{y_2 - y_1}{x_2 - x_1} = \dfrac{-5 - 7}{4 - (-2)}$

$= \dfrac{-12}{4 + 2} = \dfrac{-12}{6} = -2$

11. The slope of a line perpendicular to the line through the given points is the negative reciprocal of $-2, \dfrac{1}{2}$.

12. $d = \sqrt{(x_2 - x_1)^2 + (y_2 - y_1)^2}$

$= \sqrt{(4 - (-2))^2 + (-5 - 7)^2}$

$= \sqrt{(4 + 2)^2 + (-12)^2}$

$= \sqrt{(6)^2 + (-12)^2} = \sqrt{36 + 144}$

$= \sqrt{180}$ or $6\sqrt{5}$

13. midpoint $= \left(\dfrac{x_1 + x_2}{2}, \dfrac{y_1 + y_2}{2} \right)$

$= \left(\dfrac{-2 + 4}{2}, \dfrac{7 + (-5)}{2} \right)$

$= \left(\dfrac{2}{2}, \dfrac{2}{2} \right) = (1, 1)$

14. Write each equation in slope-intercept form.

$$2x - 5y + 7 = 0 \qquad -2x + 5y + 7 = 0$$
$$-5y = -2x - 7 \qquad\qquad 5y = 2x - 7$$
$$y = \frac{2}{5}x + \frac{7}{5} \qquad\qquad y = \frac{2}{5}x - \frac{7}{5}$$

The slope is $\dfrac{2}{5}$.   The slope is $\dfrac{2}{5}$.

Since the two slopes are equal, the lines are parallel.

15. Use the point-slope form with $(-2, 4)$ as the point and -3 as the slope.

$$y - y_1 = m(x - x_1)$$
$$y - 4 = -3(x - (-2))$$
$$y - 4 = -3(x + 2)$$
$$y - 4 = -3x - 6$$
$$3x + y + 2 = 0$$

16. First find the slope of the line.

$$m = \frac{y_2 - y_1}{x_2 - x_1} = \frac{-3-7}{4-(-1)} = \frac{-10}{4+1} = \frac{-10}{5} = -2$$

Now use the point-slope form with $(4, -3)$ as the point and -2 as the slope.

$$y - y_1 = m(x - x_1)$$
$$y - (-3) = -2(x - 4)$$
$$y + 3 = -2x + 8$$
$$2x + y - 5 = 0$$

17. Begin by finding the slope of $2x + y - 4 = 0$ by writing the equation in slope-intercept form. $2x + y - 4 = 0;\ y = -2x + 4.$ Thus, the slope of the given line is -2 and the slope of a line parallel to it is the same. Now use point slope form with $(4, 5)$ as the point and -2 as the slope.

$$y - y_1 = m(x - x_1)$$
$$y - 5 = -2(x - 4)$$
$$y - 5 = -2x + 8$$
$$2x + y - 13 = 0$$

18. Use the midpoint formula $\left(\dfrac{x_1 + x_2}{2}, \dfrac{y_1 + y_2}{2} \right)$

where (x, y) is the other endpoint.

$$\frac{5 + x}{2} = 0 \qquad \frac{-6 + y}{2} = 0$$
$$5 + x = 0 \qquad -6 + y = 0$$
$$x = -5 \qquad y = 6$$

The other endpoint is $(-5, 6)$.

19. $(-c, d); (c, d); (-a, 0); (a, 0)$

20. Let $M =$ midpoint of $\overline{AD}$.

$$M = \left(\frac{2a + 0}{2}, \frac{2b + 0}{2} \right) = (a, b)$$

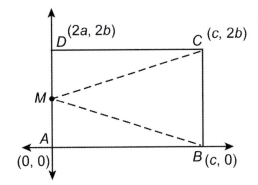

Use the distance formula to show $MB = MC$

$$MC = \sqrt{(a - c)^2 + (b - 2b)^2}$$
$$= \sqrt{(a - c)^2 + b^2}$$
$$MB = \sqrt{(a - c)^2 + (b - 0)^2}$$
$$= \sqrt{(a - c)^2 + b^2}$$

Since $MB = MC, \triangle BMC$ is isosceles.

CHAPTER 10 INTRODUCTION TO TRIGONOMETRY

Section 10.1 Sine and Cosine Ratio

10.1 PRACTICE EXERCISES

1. $\sin A = \dfrac{\text{opposite leg}}{\text{hypotenuse}} = \dfrac{8}{17}$ $\cos A = \dfrac{\text{adjacent leg}}{\text{hypotenuse}} = \dfrac{15}{17}$

$\sin B = \dfrac{\text{opposite leg}}{\text{hypotenuse}} = \dfrac{15}{17}$ $\cos B = \dfrac{\text{adjacent leg}}{\text{hypotenuse}} = \dfrac{8}{17}$

10.1 SECTION EXERCISES

1. Use the Pythagorean Theorem to find c.

$$c^2 = a^2 + b^2$$
$$= (5)^2 + (12)^2$$
$$= 25 + 144$$
$$= 169$$
$$c = \sqrt{169} = 13$$

$\sin A = \dfrac{\text{opposite leg}}{\text{hypotenuse}} = \dfrac{a}{c} = \dfrac{5}{13}$

$\cos A = \dfrac{\text{adjacent leg}}{\text{hypotenuse}} = \dfrac{b}{c} = \dfrac{12}{13}$

$\sin B = \dfrac{\text{opposite leg}}{\text{hypotenuse}} = \dfrac{b}{c} = \dfrac{12}{13}$

$\cos B = \dfrac{\text{adjacent leg}}{\text{hypotenuse}} = \dfrac{a}{c} = \dfrac{5}{13}$

3. Use the Pythagorean Theorem to find b.

$$c^2 = a^2 + b^2$$
$$(15)^2 = (7)^2 + b^2$$
$$(15)^2 - (7)^2 = b^2$$
$$225 - 49 = b^2$$
$$176 = b^2$$

Then $b = \sqrt{176} = \sqrt{16 \cdot 11} = \sqrt{16}\sqrt{11} = 4\sqrt{11}$.

$\sin A = \dfrac{\text{opposite leg}}{\text{hypotenuse}} = \dfrac{a}{c} = \dfrac{7}{15}$

$\cos A = \dfrac{\text{adjacent leg}}{\text{hypotenuse}} = \dfrac{b}{c} = \dfrac{4\sqrt{11}}{15}$

$\sin B = \dfrac{\text{opposite leg}}{\text{hypotenuse}} = \dfrac{b}{c} = \dfrac{4\sqrt{11}}{15}$

$\cos B = \dfrac{\text{adjacent leg}}{\text{hypotenuse}} = \dfrac{a}{c} = \dfrac{7}{15}$

5. Use the Pythagorean Theorem to find c.

$$c^2 = a^2 + b^2$$
$$= (0.6)^2 + (0.8)^2$$
$$= 0.36 + 0.64 = 1$$
$$c = \sqrt{1} = 1$$

$$\sin A = \frac{\text{opposite leg}}{\text{hypotenuse}} = \frac{a}{c} = \frac{0.6}{1} = 0.6$$

$$\cos A = \frac{\text{adjacent leg}}{\text{hypotenuse}} = \frac{b}{c} = \frac{0.8}{1} = 0.8$$

$$\sin B = \frac{\text{opposite leg}}{\text{hypotenuse}} = \frac{b}{c} = \frac{0.8}{1} = 0.8$$

$$\cos B = \frac{\text{adjacent leg}}{\text{hypotenuse}} = \frac{a}{c} = \frac{0.6}{1} = 0.6$$

7. Since $\sin A = \frac{5}{7} = \frac{\text{opposite leg}}{\text{hypotenuse}}$, sketch a right

triangle in which the side opposite $\angle A$ is 5 and the hypotenuse is 7.

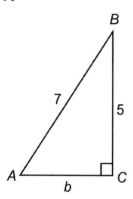

Use the Pythagorean Theorem to find b, the side adjacent to $\angle A$.

$$c^2 = a^2 + b^2$$
$$(7)^2 = (5)^2 + b^2$$
$$(7)^2 - (5)^2 = b^2$$
$$49 - 25 = b^2$$

Then $b = \sqrt{24} = \sqrt{4 \cdot 6} = \sqrt{4}\sqrt{6} = 2\sqrt{6}$.

$$\cos A = \frac{\text{adjacent leg}}{\text{hypotenuse}} = \frac{b}{c} = \frac{2\sqrt{6}}{7}$$

9. Since $\cos A = \frac{\sqrt{33}}{7} = \frac{\text{adjacent leg}}{\text{hypotenuse}}$, sketch a

right triangle in which the side adjacent to $\angle A$ is $\sqrt{33}$ and the hypotenuse is 7.

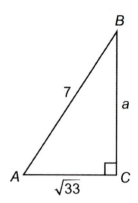

Use the Pythagorean Theorem to find a, the side opposite $\angle A$.

$$c^2 = a^2 + b^2$$
$$(7)^2 = a^2 + \left(\sqrt{33}\right)^2$$
$$(7)^2 - \left(\sqrt{33}\right)^2 = a^2$$
$$49 - 33 = a^2$$
$$16 = a^2$$

Then $a = \sqrt{16} = 4$.

$$\sin A = \frac{\text{opposite leg}}{\text{hypotenuse}} = \frac{a}{c} = \frac{4}{7}$$

11. The keystrokes for a scientific calculator are:

$$\boxed{65}\ \boxed{\sin} \rightarrow \boxed{0.906307...}$$

Thus, correct to four decimal places, $\sin 65° \approx 0.9063$. Don't forget to set the calculator in degree mode!

13. The keystrokes for a scientific calculator are:

$$\boxed{38}\ \boxed{\cos} \rightarrow \boxed{0.788010...}$$

Thus, correct to four decimal places, $\cos 38° \approx 0.7880$.

15. The keystrokes for a scientific calculator are:

$$\boxed{83.8}\ \boxed{\cos} \rightarrow \boxed{0.107999...}$$

Thus, correct to four decimal places,
$\cos 83.8° \approx 0.1080$.

17.
$$\sin 51° = \frac{\text{opposite leg}}{\text{hypotenuse}}$$

$$= \frac{x}{19} \quad \text{(multiply both sides by 19)}$$

$$19\sin 57° = x$$

$$x \approx 14.8$$

The other acute angle of the right triangle measures $39°$ $(90° - 51°)$.

$$\cos 39° = \frac{\text{adjacent leg}}{\text{hypotenuse}}$$

$$= \frac{x}{19} \quad \text{(multiply both sides by 19)}$$

$$19\cos 39° = x$$

$$x \approx 14.8$$

19.
$$\cos 45° = \frac{\text{adjacent leg}}{\text{hypotenuse}}$$

$$= \frac{x}{21} \quad \text{(multiply both sides by 21)}$$

$$21\cos 45° = x$$

$$x \approx 14.8$$

The other acute angle of the right triangle measures $45°$ $(90° - 45°)$.

$$\sin 45° = \frac{\text{opposite leg}}{\text{hypotenuse}}$$

$$= \frac{x}{21} \quad \text{(multiply both sides by 21)}$$

$$21\sin 45° = x$$

$$x \approx 14.8$$

21.
$$\sin 37° = \frac{\text{opposite leg}}{\text{hypotenuse}}$$

$$= \frac{x}{35} \quad \text{(multiply both sides by 35)}$$

$$35\sin 37° = x$$

$$x \approx 21.1$$

$$\cos 37° = \frac{\text{adjacent leg}}{\text{hypotenuse}}$$

$$= \frac{y}{35} \quad \text{(multiply both sides by 35)}$$

$$35\cos 37° = y$$

$$y \approx 28.0$$

23.
$$\cos 61° = \frac{\text{adjacent leg}}{\text{hypotenuse}}$$

$$= \frac{x}{20} \quad \text{(multiply both sides by 20)}$$

$$20\cos 61° = x$$

$$x \approx 9.7$$

$$\sin 61° = \frac{\text{opposite leg}}{\text{hypotenuse}}$$

$$= \frac{y}{20} \quad \text{(multiply both sides by 20)}$$

$$20\sin 61° = y$$

$$y \approx 17.5$$

25. If the given leg is adjacent to the given acute angle, use the cosine ratio. If the given leg is opposite the given acute angle, use the sine ratio.

Section 10.2 Tangent Ratio

10.2 PRACTICE EXERCISES

1. $\tan A = \dfrac{\text{opposite leg}}{\text{adjacent leg}} = \dfrac{5}{12}$

$\tan B = \dfrac{\text{opposite leg}}{\text{adjacent leg}} = \dfrac{12}{5}$

2. The keystrokes for a scientific calculator are:

$\boxed{0.90630779}\ \boxed{\cos^{-1}} \to \boxed{24.999984...}$

Thus, correct to the nearest degree, $A \approx 25°$.

10.2 SECTION EXERCISES

1. $\tan A = \dfrac{\text{opposite leg}}{\text{adjacent leg}} = \dfrac{a}{b} = \dfrac{5}{12}$

$\tan B = \dfrac{\text{opposite leg}}{\text{adjacent leg}} = \dfrac{b}{a} = \dfrac{12}{5}$

3. Use the Pythagorean Theorem to find b.

$$a^2 + b^2 = c^2$$
$$7^2 + b^2 = 15^2$$
$$49 + b^2 = 225$$
$$b^2 = 176$$
$$b = \sqrt{176} = \sqrt{16 \cdot 11}$$
$$= 4\sqrt{11}$$

$\tan A = \dfrac{\text{opposite leg}}{\text{adjacent leg}} = \dfrac{a}{b}$

$= \dfrac{7}{4\sqrt{11}}$ or $\dfrac{7}{4\sqrt{11}} \cdot \dfrac{\sqrt{11}}{\sqrt{11}} = \dfrac{7\sqrt{11}}{44}$

$\tan B = \dfrac{\text{opposite leg}}{\text{adjacent leg}} = \dfrac{b}{a} = \dfrac{4\sqrt{11}}{7}$

5. $\tan A = \dfrac{\text{opposite leg}}{\text{adjacent leg}} = \dfrac{a}{b} = \dfrac{0.6}{0.8} = \dfrac{3}{4}$

$\tan B = \dfrac{\text{opposite leg}}{\text{adjacent leg}} = \dfrac{b}{a} = \dfrac{0.8}{0.6} = \dfrac{4}{3}$

7. Since $\tan A = \dfrac{\sqrt{5}}{2} = \dfrac{\text{opposite leg}}{\text{adjacent leg}}$, the leg opposite $\angle A$ is $\sqrt{5}$ and the leg adjacent to $\angle A$ is 2.

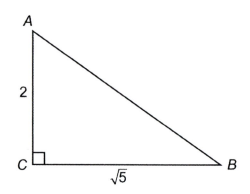

Use the Pythagorean Theorem to find c.

$$a^2 + b^2 = c^2$$
$$\left(\sqrt{5}\right)^2 + 2^2 = c^2$$
$$5 + 4 = c^2$$
$$9 = c^2$$
$$3 = c$$

$\sin A = \dfrac{\text{opposite leg}}{\text{hypotenuse}} = \dfrac{a}{c} = \dfrac{\sqrt{5}}{3}$

$\cos A = \dfrac{\text{adjacent leg}}{\text{hypotenuse}} = \dfrac{b}{c} = \dfrac{2}{3}$

9. Since $\tan A = 7 = \frac{7}{1} = \frac{\text{opposite leg}}{\text{adjacent leg}}$, sketch a

triangle in which the side opposite $\angle A$ is 7 and the side adjacent to $\angle A$ is 1.

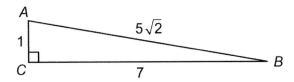

Use the Pythagorean Theorem to find c, the hypotenuse.

$$c^2 = a^2 + b^2$$
$$= (7)^2 + (1)^2$$
$$= 49 + 1$$
$$= 50$$

Then $c = \sqrt{50} = \sqrt{25 \cdot 2} = \sqrt{25}\sqrt{2} = 5\sqrt{2}$.

$$\sin A = \frac{\text{opposite leg}}{\text{hypotenuse}} = \frac{a}{c}$$

$$= \frac{7}{5\sqrt{2}} \text{ or } \frac{7\sqrt{2}}{5\sqrt{2}\sqrt{2}}$$

$$= \frac{7\sqrt{2}}{5(2)} = \frac{7\sqrt{2}}{10}$$

$$\cos A = \frac{\text{adjacent leg}}{\text{hypotenuse}} = \frac{b}{c}$$

$$= \frac{1}{5\sqrt{2}} \text{ or } \frac{1 \cdot \sqrt{2}}{5\sqrt{2}\sqrt{2}}$$

$$= \frac{\sqrt{2}}{5(2)} = \frac{\sqrt{2}}{10}$$

11. The keystrokes for a scientific calculator are:

$$\boxed{38}\ \boxed{\tan} \rightarrow \boxed{0.781285...}$$

Thus, rounded to 4 decimal places

$$\tan 38° \approx 0.7813.$$

13. The keystrokes for a scientific calculator are:

$$\boxed{81.43}\ \boxed{\tan} \rightarrow \boxed{6.635689...}$$

Thus, rounded to 4 decimal places

$$\tan 81.43° \approx 6.6357.$$

15. To find $m\angle A$, use these keystrokes for a scientific calculator:

$$\boxed{0.1258}\ \boxed{\sin^{-1}} \rightarrow \boxed{7.226957...}$$

Thus, to the nearest tenth of a degree,

$$m\angle A \approx 7.2°.$$

17. To find $m\angle A$, use these keystrokes for a scientific calculator:

$$\boxed{0.9301}\ \boxed{\tan^{-1}} \rightarrow \boxed{42.925897...}$$

Thus, to the nearest tenth of a degree,

$$m\angle A \approx 42.9°.$$

19. To find $m\angle A$, use the following keystrokes for a scientific calculator:

$$\boxed{0.1022}\ \boxed{\cos^{-1}} \rightarrow \boxed{84.134129...}$$

Thus, to the nearest tenth of a degree,

$$m\angle A \approx 84.1°.$$

21. $\tan 27° = \dfrac{\text{opposite leg}}{\text{adjacent leg}} = \dfrac{b}{10}$

$$10 \tan 27° = b$$
$$b \approx 5.1$$

23. $\tan x = \dfrac{\text{opposite leg}}{\text{adjacent leg}} = \dfrac{3}{4}$

$$\tan x = \frac{3}{4}$$

$$\boxed{\frac{3}{4}}\ \boxed{=}\ \boxed{\tan^{-1}} \rightarrow \boxed{36.869897...}$$

Thus, to the nearest tenth of a degree, the measure of the desired angle is approximately 36.9°.

25. $\sin x = \dfrac{\text{opposite leg}}{\text{hypotenuse}} = \dfrac{10}{26}$

$\sin x = \dfrac{10}{26}$

$\boxed{\dfrac{10}{26}} = \boxed{\sin^{-1}} \rightarrow \boxed{22.619864...}$

Thus, to the nearest tenth of a degree, the measure of the desired angle is approximately 22.6°.

27. $\tan x = \dfrac{\text{opposite leg}}{\text{adjacent leg}} = \dfrac{2}{\sqrt{5}}$

$\tan x = \dfrac{2}{\sqrt{5}}$

$\boxed{\dfrac{2}{\sqrt{5}}} = \boxed{\tan^{-1}} \rightarrow \boxed{41.810314...}$

Thus, to the nearest tenth of a degree, the measure of the desired angle is approximately 41.8°.

29. $\tan x = \dfrac{\text{opposite leg}}{\text{adjacent leg}} = \dfrac{10}{10\sqrt{3}}$

$\tan x = \dfrac{10}{10\sqrt{3}}$

$\boxed{\dfrac{10}{10\sqrt{3}}} = \boxed{\tan^{-1}} \rightarrow \boxed{30}$

Thus, to the nearest tenth of a degree, the measure of the desired angle is 30°.

31. $\cos x = \dfrac{\text{adjacent leg}}{\text{hypotenuse}} = \dfrac{8}{16}$

$\cos x = \dfrac{8}{16}$

$\boxed{\dfrac{8}{16}} = \boxed{\cos^{-1}} \rightarrow \boxed{60}$

Thus the measure of the desired angle is 60°.

33. The diameter of the circle is also the hypotenuse of the inscribed right triangle. $d = 2r$ thus $d = 2(14) = 28$ inches.

$$\sin 26° = \dfrac{\text{opposite leg}}{\text{hypotenuse}} = \dfrac{BC}{28}$$

$$28\sin 26° = BC$$

$$BC \approx 12.3 \text{ in}$$

35. (a) $m = \dfrac{4-0}{5-0} = \dfrac{4}{5}$

(b) To find tan A, the length of the opposite leg (BC) and the adjacent leg (AB) are needed.

$$BC = \sqrt{(5-5)^2 + (4-0)^2} = \sqrt{0+4^2} = 4$$
$$AB = \sqrt{(5-0)^2 + (0-0)^2} = \sqrt{5^2+0} = 5$$
$$\tan A = \dfrac{\text{opposite leg}}{\text{adjacent leg}} = \dfrac{4}{5}$$

(c) Yes, it will always be true. The slope of $\overline{AC}$ = tan A because the slope of a line is the change in y divided by the change in x:

$$\dfrac{y_2 - y_1}{x_2 - x_1}.$$

The length of the opposite leg is the change in y, while the length of the adjacent leg is the change in x. Thus,

$$\dfrac{y_2 - y_1}{x_2 - x_1} = \dfrac{\text{opposite leg}}{\text{adjacent leg}}.$$

Section 10.3 Solving Right Triangles

10.3 PRACTICE EXERCISE

1. First sketch a right triangle in which $b = 10$ and $c = 16$ like the one below.

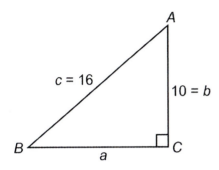

To find a: Use the Pythagorean Theorem.

$$a^2 + b^2 = c^2$$
$$a^2 + (10)^2 = (16)^2$$
$$a^2 = (16)^2 - (10)^2$$
$$a^2 = 256 - 100 = 156$$
$$a = \sqrt{156} \approx 12.489996$$

Thus, to the nearest unit, $a \approx 12$.

To find $m\angle A$ (use the given sides b and c):

$$\cos A = \frac{b}{c} = \frac{10}{16} = 0.625$$

Use these keystrokes on a scientific calculator.

$$\boxed{0.625}\ \boxed{\cos^{-1}} \rightarrow \boxed{51.317812...}$$

Thus, $m\angle A \approx 51°$.

To find $m\angle B$: Since $\angle A$ and $\angle B$ are complementary,

$$m\angle B = 90° - m\angle A$$
$$\approx 90° - 51°$$
$$\approx 39°.$$

Thus $m\angle A \approx 51°, m\angle B \approx 39°$ and $a \approx 12$

10.3 SECTION EXERCISES

1. First sketch a right triangle in which $a = 9$ and $m\angle A = 60°$.

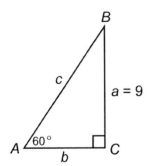

Since $\angle A$ and $\angle B$ are complementary,

$$m\angle A + m\angle B = 90°$$
$$60° + m\angle B = 90°$$
$$m\angle B = 30°.$$

To find b we use the given information.

$$\tan A = \frac{a}{b}$$
$$\tan 60° = \frac{9}{b}$$
$$b \tan 60° = 9$$
$$b = \frac{9}{\tan 60°}$$

The keystrokes for a scientific calculator are:

$\boxed{9}\ \boxed{\div}\ \boxed{60}\ \boxed{\tan}\ \boxed{=} \rightarrow \boxed{5.196152...}$

Remember to set the calculator in degree mode. Round to the nearest whole number, $b \approx 5$.

To find c use the given information.

$$\sin A = \frac{a}{c}$$

$$\sin 60° = \frac{9}{c}$$

$$c \sin 60° = 9$$

$$c = \frac{9}{\sin 60°}$$

The keystrokes for a scientific calculator are:

$\boxed{9}\ \boxed{\div}\ \boxed{60}\ \boxed{\sin}\ \boxed{=} \rightarrow \boxed{10.392304...}$

Then to the nearest whole number, $c \approx 10$. Thus, the missing parts of the triangle are $m\angle B = 30°$, $b \approx 5$, and $c \approx 10$.

3. First sketch a right triangle in which $c = 12$ and $m\angle A = 20°$.

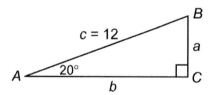

Since $\angle A$ and $\angle B$ are complementary,

$$m\angle A + m\angle B = 90°$$

$$20° + m\angle B = 90°$$

$$m\angle B = 70°.$$

Using the given information:

$$\sin A = \frac{a}{c}$$

$$\sin 20° = \frac{a}{12}$$

$$12 \sin 20° = a$$

The keystrokes for a scientific calculator are:

$\boxed{12}\ \boxed{\times}\ \boxed{20}\ \boxed{\sin}\ \boxed{=} \rightarrow \boxed{4.104241...}$

To the nearest whole number, $a \approx 4$.

Using the given information:

$$\cos A = \frac{b}{c}$$

$$\cos 20° = \frac{b}{12}$$

$$12 \cos 20° = b$$

The keystrokes for a scientific calculator are:

$\boxed{12}\ \boxed{\times}\ \boxed{20}\ \boxed{\cos}\ \boxed{=} \rightarrow \boxed{11.276311...}$

Then to the nearest whole number, $b \approx 11$. Thus, the missing parts of the triangle are $m\angle B = 70°$, $a \approx 4$, and $b \approx 11$.

5. First sketch the right triangle in which $a = 12$ and $c = 13$.

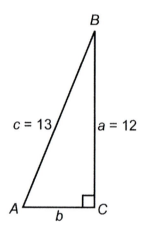

To find b we use the Pythagorean Theorem.

$$c^2 = a^2 + b^2$$

$$(13)^2 = (12)^2 + b^2$$

$$169 = 144 + b^2$$

$$25 = b^2$$

$$5 = b$$

Using the given information:

$$\sin A = \frac{a}{c}$$

$$\sin A = \frac{12}{13}$$

The keystrokes for a scientific calculator are:

$\boxed{12} \boxed{\div} \boxed{13} \boxed{=} \boxed{\sin^{-1}} \rightarrow \boxed{67.380135...}$

Then $m\angle A \approx 67°$, correct to the nearest degree.

To find $m\angle B$ we could subtract $m\angle A$ from $90°$ since $\angle A$ and $\angle B$ are complementary. This would give $m\angle B \approx 23°$. We could also note that

$$\cos B = \frac{12}{13}.$$

The keystrokes for a scientific calculator are:

$\boxed{12} \boxed{\div} \boxed{13} \boxed{=} \boxed{\cos^{-1}} \rightarrow \boxed{22.619864...}$

Then in either case, $m\angle B \approx 23°$, correct to the nearest degree. Thus, the missing parts of the triangle are $b = 5$, $m\angle A \approx 67°$, and $m\angle B \approx 23°$.

7. First sketch the right triangle in which $b = 7$ and $c = 15$.

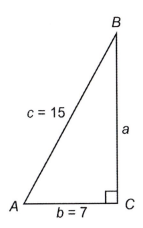

To find a: By the Pythagorean Theorem we have

$$c^2 = a^2 + b^2$$
$$(15)^2 = a^2 + (7)^2$$
$$225 = a^2 + 49$$
$$176 = a^2$$
$$a = \sqrt{176} \approx 13.266499...$$

Then to the nearest whole number, $a = 13$.

To find $m\angle A$: $\cos A = \dfrac{7}{15}$

$\boxed{7} \boxed{\div} \boxed{15} \boxed{=} \boxed{\cos^{-1}} \rightarrow \boxed{62.181861...}$

Then to the nearest degree, $m\angle A \approx 62°$.

To find $\angle B$: $\sin B = \dfrac{7}{15}$

$\boxed{7} \boxed{\div} \boxed{15} \boxed{=} \boxed{\sin^{-1}} \rightarrow \boxed{27.818139...}$

Then to the nearest degree, $m\angle B \approx 28°$.

Note that we could also have found $m\angle B$ by subtracting $62°$ from $90°$, but if $m\angle A$ were incorrect, we would have compounded the error. In fact, we can use this observation as a check.

Thus $a \approx 13$, $m\angle B \approx 28°$, $m\angle A \approx 62°$.

9. First sketch a right triangle in which $a = 9.2$ and $c = 10.1$.

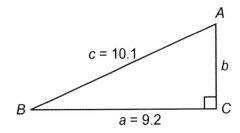

To find b: Use the Pythagorean Theorem.

$$c^2 = a^2 + b^2$$
$$(10.1)^2 = (9.2)^2 + b^2$$
$$102.1 = 84.64 + b^2$$
$$17.37 = b^2$$
$$4.1677332 \approx b$$

Then to the nearest tenth, $b \approx 4.2$.

To find $m\angle A$: $\sin A = \dfrac{9.2}{10.1}$

$\boxed{9.2} \boxed{\div} \boxed{10.1} \boxed{=} \boxed{\sin^{-1}} \rightarrow \boxed{65.628785...}$

Then to the nearest degree, $m\angle A \approx 65.6°$.

To find $m\angle B$: $\cos B = \dfrac{9.2}{10.1}$

$\boxed{9.2}\;\boxed{\div}\;\boxed{10.1}\;\boxed{=}\;\boxed{\sin^{-1}}\rightarrow\boxed{24.371214...}$

Then to the nearest tenth of a degree,
$m\angle B \approx 24.4°$.

Thus $b \approx 4.2$, $m\angle A \approx 65.6°$, $m\angle B \approx 24.4°$.

11. Sketch a right triangle in which $m\angle A = 26.7°$ and $c = 12.0$.

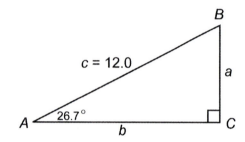

To find $m\angle B$:

$$m\angle A + m\angle B = 90°$$
$$26.7° + m\angle B = 90°$$
$$m\angle B = 90° - 26.7°$$
$$m\angle B = 63.3°$$

To find a: $\sin 26.7° = \dfrac{a}{12.0}$

$$12.0\sin 26.7° = a$$

$\boxed{12.0}\;\boxed{\times}\;\boxed{26.7}\;\boxed{\cos^{-1}}\;\boxed{=}\rightarrow\boxed{5.391828...}$

Then to the nearest tenth, $a \approx 5.4$.

To find b: $\cos 26.7° = \dfrac{b}{12.0}$

$$12.0\cos 26.7° = b$$

$\boxed{12.0}\;\boxed{\times}\;\boxed{26.7}\;\boxed{\cos}\;\boxed{=}\rightarrow\boxed{10.720457...}$

Then to the nearest tenth, $b \approx 10.7$.

Thus $a \approx 5.4$, $b \approx 10.7$, $m\angle B = 63.3°$.

13. Sketch a right triangle in which $b = 3.2$ and $m\angle B = 10.8°$.

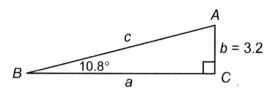

To find $m\angle A$:

$$m\angle A + m\angle B = 90°$$
$$m\angle A + 10.8° = 90°$$
$$m\angle A = 79.2°$$

To find a: $\tan 10.8° = \dfrac{3.2}{a}$

$$a\tan 10.8° = 3.2$$

$$a = \dfrac{3.2}{\tan 10.8°}$$

$\boxed{3.2}\;\boxed{\div}\;\boxed{10.8}\;\boxed{\tan}\;\boxed{=}\rightarrow\boxed{16.774987...}$

Then to the nearest tenth, $a \approx 16.8$.

To find c: $\sin 10.8° = \dfrac{3.2}{c}$

$$c\sin 10.8° = 3.2$$

$$c = \dfrac{3.2}{\sin 10.8°}$$

$\boxed{3.2}\;\boxed{\div}\;\boxed{10.8}\;\boxed{\sin}\;\boxed{=}\rightarrow\boxed{17.077476...}$

Then to the nearest tenth, $c \approx 17.1$.

Thus $a \approx 16.8$, $c \approx 17.1$, $m\angle A = 79.2°$.

15. Sketch a right triangle in which $m\angle A = 22.5°$ and $c = 28.7$.

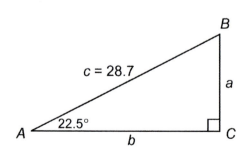

To find $m\angle B$:

$$m\angle A + m\angle B = 90°$$
$$22.5° + m\angle B = 90°$$
$$m\angle B = 67.5°$$

To find a: $\sin 22.5° = \dfrac{a}{28.7}$

$$28.7\sin 22.5° = a$$

$\boxed{28.7}\ \boxed{\times}\ \boxed{22.5}\ \boxed{\sin}\ \boxed{=} \rightarrow \boxed{10.983014...}$

Then to the nearest tenth, $a \approx 11.0$.

To find b: $\cos 22.5° = \dfrac{b}{28.7}$

$$28.7\cos 22.5° = b$$

$\boxed{28.7}\ \boxed{\times}\ \boxed{22.5}\ \boxed{\cos}\ \boxed{=} \rightarrow \boxed{26.515343...}$

Then rounded to the nearest tenth, $b \approx 26.5$.
Thus $a \approx 11.0$, $b \approx 26.5$, $m\angle B = 67.5°$.

17. Sketch a right triangle in which $c = 21.9$ and $m\angle B = 81.7°$.

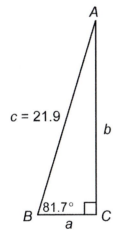

To find $m\angle A$:

$$m\angle A + m\angle B = 90°$$
$$m\angle A + 81.7° = 90°$$
$$m\angle A = 8.3°$$

To find a: $\cos 81.7° = \dfrac{a}{21.9}$

$$21.9\cos 81.7° = a$$

$\boxed{21.9}\ \boxed{\times}\ \boxed{81.7}\ \boxed{\cos}\ \boxed{=} \rightarrow \boxed{3.161400...}$

Then to the nearest tenth, $a \approx 3.2$.

To find b: $\sin 81.7° = \dfrac{b}{21.9}$

$$21.9\sin 81.7° = b$$

$\boxed{21.9}\ \boxed{\times}\ \boxed{81.7}\ \boxed{\sin}\ \boxed{=} \rightarrow \boxed{21.670615...}$

Then to the nearest tenth, $b \approx 21.7$.

Thus $a \approx 3.2$, $b \approx 21.7$, $m\angle A = 8.3°$.

19. Sketch a rectangle similar to the one in the problem.

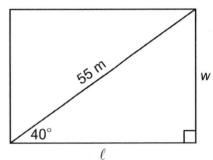

Let w = width and ℓ = length of the rectangle.
To find w, use the sine ratio.

$$\sin 40° = \frac{w}{55}$$
$$55\sin 40° = w$$

To find ℓ, use the cosine ratio

$$\cos 40° = \frac{\ell}{55}$$
$$55\cos 40° = \ell$$

The perimeter of a rectangle is $P = 2\ell + 2w$.

$$P = 2(55\cos 40°) + 2(55\sin 40°)$$

$$P \approx 155.0 \text{ m (rounded to the nearest tenth)}$$

21. Sketch a triangle similar to the one in the problem.

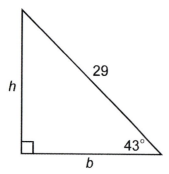

Let h = height and b = base of the triangle.
To find the height of the triangle, use the sine ratio.

$$\sin 43° = \frac{h}{29}$$

$$29\sin 43° = h$$

To find the base of the triangle, use the cosine ratio.

$$\cos 43° = \frac{b}{29}$$

$$29\cos 43° = b$$

The area of a triangle is $A = \frac{1}{2}bh$.

$$A = \frac{1}{2}\left(29\cos 43°\right)\left(29\sin 43°\right)$$

$$A \approx 209.7 \text{ cm (rounded to the nearest tenth)}$$

23. Sketch a pentagon with the measurements given in the problem.

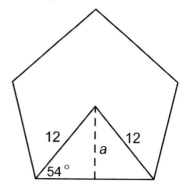

Each central angle measures $\frac{360}{5} = 72°$.

The triangle drawn in the figure is isosceles therefore the other two angles of the triangle measure $\frac{180-72}{2} = \frac{108}{2} = 54°$.

Let a = length of the apothem.
Use the sine ratio to find the apothem.

$$\sin 54° = \frac{a}{12}$$

$$12\sin 54° = a$$

Thus, the length of the apothem is approximately 9.7 in (rounded to the nearest tenth).

Section 10.4 Applications Involving Right Triangles

10.4 PRACTICE EXERCISE

1. Use Figure 10.20. What we must find is *BF*.

$$\cos B = \frac{AB}{BF}$$

$$\cos 38.8° = \frac{11.0}{BF}$$

$$BF \cos 38.8° = 11.0$$

$$BF = \frac{11.0}{\cos 38.8°}$$

$\boxed{11.0}\ \boxed{\div}\ \boxed{38.8}\ \boxed{\cos}\ \boxed{=} \rightarrow \boxed{14.114544}$

Thus, to the nearest tenth of a mile, the fire is about 14.1 miles from the second lookout.

10.4 SECTION EXERCISES

1. ∠3 is an angle of elevation.

3. ∠1 is an angle of depression.

5. Start by drawing a picture.

Let x = distance from the base of the cliff to the ship.

$$\tan 10° = \frac{250}{x}$$
$$x \tan 10° = 250$$
$$x = \frac{250}{\tan 10°}$$

The keystrokes on a scientific calculator are:

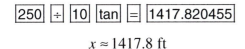

$$x \approx 1417.8 \text{ ft}$$

It is approximately 1417.8 ft from the ship to the base of the cliff.

7. Start by drawing a picture.

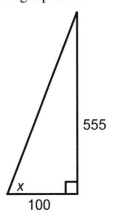

Let x = the angle of elevation.

$$\tan x = \frac{555}{100}$$

The keystrokes on a scientific calculator are:

$$\boxed{\frac{555}{100}} \boxed{=} \boxed{\tan^{-1}} \rightarrow \boxed{79.786027...}$$

Thus, the angle of elevation, measured to the nearest tenth of a degree is approximately 79.8°.

9. From the information and the figure given in the text, if x is the height of the pole, then

$$\tan 49° = \frac{x}{40}$$
$$40 \tan 49° = x$$

The keystrokes on a scientific calculator are:

$$\boxed{40} \boxed{\times} \boxed{49} \boxed{\tan} \boxed{=} \rightarrow \boxed{46.014736...}$$

Thus, to the nearest tenth of a foot, the pole is approximately 46.0 ft tall.

11. Make a sketch using the information given.

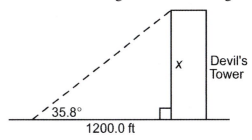

If x is the height of Devil's Tower, then

$$\tan 35.8 = \frac{x}{1200.0}$$
$$1200.0 \tan 35.8 = x$$

The keystrokes on a scientific calculator are:

$$\boxed{1200.0} \boxed{\times} \boxed{35.8} \boxed{\tan} \boxed{=} \rightarrow \boxed{865.467295...}$$

Thus, to the nearest tenth of a foot, the height of the tower is approximately 865.5 ft.

13. Make a sketch using the information given.

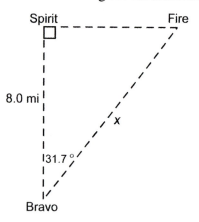

We must find the value of x.

$$\cos 31.7° = \frac{8.0}{x}$$

$$x\cos 31.7° = 8.0$$

$$x = \frac{8.0}{\cos 31.7°}$$

The keystrokes on a scientific calculator are:

$$\boxed{8.0} \; \boxed{÷} \; \boxed{31.7} \; \boxed{\cos} \; \boxed{=} \; → \; \boxed{9.402792...}$$

Thus, to the nearest tenth of a mile, the fire is approximately 9.4 mi from Lookout Bravo.

15. Make a sketch using the information given.

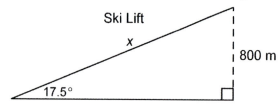

We must find the value of x.

$$\sin 17.5° = \frac{800}{x}$$

$$x\sin 17.5° = 800$$

$$x = \frac{800}{\sin 17.5°}$$

The keystrokes on a scientific calculator are:

$$\boxed{800} \; \boxed{÷} \; \boxed{17.5} \; \boxed{\sin} \; \boxed{=} \; → \; \boxed{2660.407619}$$

Thus, to the nearest tenth of a meter, the length of the ski lift ride is approximately 2660.4 m.

17. First sketch an isosceles triangle with vertex angle 90° and equal sides 82 ft.

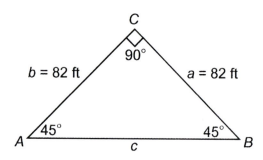

Since the triangle is isosceles, $m\angle A = m\angle B = 45°$. To find the perimeter we must find c. Relative to $\angle A$.

$$\sin 45° = \frac{82}{c}$$

$$c\sin 45° = 82$$

$$c = \frac{82}{\sin 45°}$$

(a) The perimeter of the triangle is

$$82 + 82 + \frac{82}{\sin 45°} \approx 279.965512...$$

Thus the perimeter, rounded to the nearest foot is approximately 280 ft.

(b) The area of the triangle is $\frac{1}{2}$(base)(height). We can use $a = 82$ for the base and $b = 82$ for the height. Thus, the area is

$$\frac{1}{2}(82)(82) = (41)(82) = 3362 \text{ ft}^2.$$

19. First sketch an equilateral triangle with sides 6.2 mi. Let x be the height of the triangle.

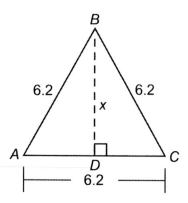

Since $m\angle A = m\angle B = m\angle C = 60°$, relative to $\triangle ABD$,

$$\sin A = \frac{x}{6.2}$$

$$\sin 60° = \frac{x}{6.2}$$

$$6.2 \sin 60° = x$$

(a) The perimeter of the triangle is

$$6.2 + 6.2 + 6.2 = 18.6 \text{ mi.}$$

(b) The area of the triangle is $\frac{1}{2}$(base)(height). Then the base is 6.2 and the height is $6.2 \sin 60°$.

$$A = \frac{1}{2}(6.2)(6.2 \sin 60°)$$

$$\boxed{1}\,\boxed{÷}\,\boxed{2}\,\boxed{×}\,\boxed{6.2}\,\boxed{×}\,\boxed{6.2}\,\boxed{×}\,\boxed{60}\,\boxed{\sin}\,\boxed{=}$$

$$\rightarrow \boxed{16.645008...}$$

Thus, correct to the nearest tenth of a square mile, the area is approximately 16.6 mi^2.

Chapter 10 Review Exercises

1. Make a sketch of $\triangle ABC$ with $a = 16$ and $b = 12$.

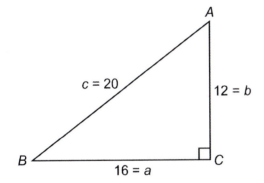

Use the Pythagorean theorem to find the hypotenuse c.

$$c^2 = a^2 + b^2$$

$$c^2 = (16)^2 + (12)^2 = 256 + 144$$

$$= 400$$

$$c = \sqrt{400} = 20$$

$$\sin A = \frac{\text{opposite side}}{\text{hypotenuse}} = \frac{a}{c} = \frac{16}{20} = \frac{4}{5}$$

$$\cos A = \frac{\text{adjacent side}}{\text{hypotenuse}} = \frac{b}{c} = \frac{12}{20} = \frac{3}{5}$$

$$\sin B = \frac{\text{opposite side}}{\text{hypotenuse}} = \frac{b}{c} = \frac{12}{20} = \frac{3}{5}$$

$$\cos B = \frac{\text{adjacent side}}{\text{hypotenuse}} = \frac{a}{c} = \frac{16}{20} = \frac{4}{5}$$

2. Make a sketch of $\triangle ABC$ with $a = 15$ and $c = 20$.

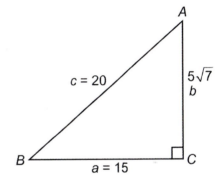

Use the Pythagorean theorem to find b.

$$c^2 = a^2 + b^2$$

$$(20)^2 = (15)^2 + b^2$$

Then $\quad b^2 = (20)^2 - (15)^2$

$$= 400 - 225 = 175$$

$$b = \sqrt{175} = \sqrt{25 \cdot 7} = 5\sqrt{7}$$

$$\sin A = \frac{a}{c} = \frac{15}{20} = \frac{3}{4}$$

$$\cos A = \frac{b}{c} = \frac{5\sqrt{7}}{20} = \frac{\sqrt{7}}{4}$$

$$\sin B = \frac{b}{c} = \frac{5\sqrt{7}}{20} = \frac{\sqrt{7}}{4}$$

$$\cos B = \frac{a}{c} = \frac{15}{20} = \frac{3}{4}$$

3. Sketch a right triangle with $\cos B = \frac{1}{3}$; that is, the side adjacent to $\angle B = 1$ and the hypotenuse = 3.

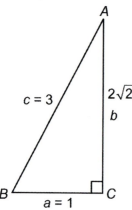

Use the Pythagorean Theorem to find b.

$$c^2 = a^2 + b^2$$
$$(3)^2 = (1)^2 + b^2$$
$$9 = 1 + b^2$$
$$8 = b^2$$

Thus, $b = \sqrt{8} = \sqrt{4 \cdot 2} = \sqrt{4}\sqrt{2} = 2\sqrt{2}$

$$\sin B = \frac{b}{c} = \frac{2\sqrt{2}}{3}$$

4. To find $\sin 21°$, use:

$$\boxed{21}\ \boxed{\sin} \rightarrow \boxed{0.358368...}$$

To four decimal places, $\sin 21° \approx 0.3584$.

5. To find $\cos 65.8°$, use:

$$\boxed{65.8}\ \boxed{\cos} \rightarrow \boxed{0.409923...}$$

To four decimal places, $\cos 65.8° \approx 0.4099$.

6. $\sin 38° = \dfrac{\text{opposite leg}}{\text{hypotenuse}} = \dfrac{x}{43}$

$43 \sin 38° = x$

The keystrokes on a scientific calculator are:

$$\boxed{43}\ \boxed{\times}\ \boxed{38}\ \boxed{\sin}\ \boxed{=} \rightarrow \boxed{26.473443...}$$

To one decimal place, $x \approx 26.5$.

$\cos 38° = \dfrac{\text{adjacent leg}}{\text{hypotenuse}} = \dfrac{y}{43}$

$43 \cos 38° = y$

The keystrokes on a scientific calculator are:

$$\boxed{43}\ \boxed{\times}\ \boxed{38}\ \boxed{\cos}\ \boxed{=} \rightarrow \boxed{33.884462...}$$

To one decimal place, $y \approx 33.9$.

7. Make a sketch of the right triangle with $\tan A = \frac{20}{15}$; that is, the side opposite to $\angle A$ is 20 and the side adjacent to it is 15.

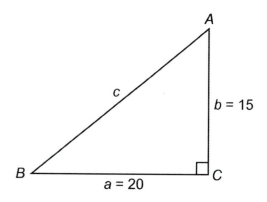

Find c using the Pythagorean Theorem.

$$c^2 = a^2 + b^2$$
$$c^2 = 20^2 + 15^2$$
$$c^2 = 400 + 225 = 625$$
$$c = 25$$

$$\sin A = \frac{a}{c} = \frac{20}{25} = \frac{4}{5}$$
$$\cos A = \frac{b}{c} = \frac{15}{25} = \frac{3}{5}$$

8. To find $\tan 82.1°$ use:

$$\boxed{82.1}\ \boxed{\tan} \rightarrow \boxed{7.206611...}$$

To four decimal places, $\tan 82.1° \approx 7.2066$.

9. To find $m\angle A$ use:

$$\boxed{0.2196}\ \boxed{\sin^{-1}} \rightarrow \boxed{12.685540...}$$

To the nearest tenth of a degree,
$m\angle A \approx 12.7°$.

10. To find $m\angle A$ use:

$$\boxed{0.9108}\ \boxed{\cos^{-1}} \rightarrow \boxed{24.383859...}$$

To the nearest tenth of a degree,
$m\angle A \approx 24.4°$.

11. To find $m\angle A$ use:

$$\boxed{1.6432}\ \boxed{\tan^{-1}} \rightarrow \boxed{58.676617...}$$

To the nearest tenth of a degree,
$m\angle A \approx 58.7°$.

12. $\sin x = \dfrac{\text{opposite leg}}{\text{hypotenuse}} = \dfrac{24}{30}$ or $\dfrac{4}{5}$

$$\boxed{\dfrac{4}{5}}\ \boxed{=}\ \boxed{\sin^{-1}} \rightarrow \boxed{53.130102...}$$

Thus, to the nearest degree, the measure of
the desired angle is approximately 53°.

13. $\tan y = \dfrac{\text{opposite leg}}{\text{hypotenuse}} = \dfrac{12}{12}$ or 1

$$\boxed{1}\ \boxed{\tan^{-1}} \rightarrow \boxed{45}$$

Thus the measure of the desired angle is 45°.

14. Sketch a right triangle with $b = 6$ and
$m\angle A = 30°$.

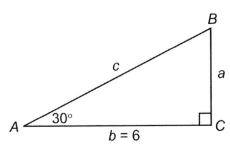

To find $\angle B$: $m\angle A + m\angle B = 90°$
$$30° + m\angle B = 90°$$
$$m\angle B = 60°$$

To find a: $\quad \tan A = \dfrac{a}{b}$

$$\tan 30° = \dfrac{a}{6}$$

$$6\tan 30° = a$$

$$\boxed{6}\ \boxed{\times}\ \boxed{30}\ \boxed{\tan}\ \boxed{=} \rightarrow \boxed{3.464101...}$$

Then $a \approx 3$, correct to the nearest whole
number.

To find c: $\quad \cos A = \dfrac{b}{c}$

$$\cos 30° = \dfrac{6}{c}$$

$$c\cos 30° = 6$$

$$c = \dfrac{6}{\cos 30°}$$

$$\boxed{6}\ \boxed{\div}\ \boxed{30}\ \boxed{\cos}\ \boxed{=} \rightarrow \boxed{6.928203...}$$

Then $c \approx 7$, correct to the nearest whole
number.

Thus, $m\angle B = 60°$, $a \approx 3$ and $c \approx 7$.

15. Sketch a right triangle with $c = 12$ and
$m\angle B = 42°$.

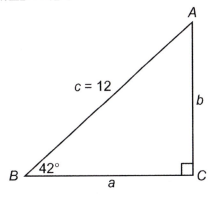

To find $\angle A$: $m\angle A + m\angle B = 90°$
$$m\angle A + 42° = 90°$$
$$m\angle A = 48°$$

To find a: $\cos 42° = \dfrac{a}{12}$

$12\cos 42° = a$

$8.917738... = a$

Then $a \approx 9$, correct to the nearest whole number.

To find b: $\sin 42° = \dfrac{b}{12}$

$12\sin 42° = b$

$8.029567... = b$

Then $b \approx 8$, correct to the nearest whole number.

Thus, $m\angle A = 48°$, $a \approx 9$ and $b \approx 8$.

16. Sketch a right triangle with $a = 2.4$ and $m\angle A = 72.3°$.

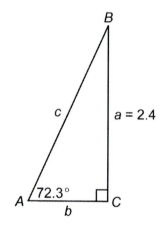

To find $\angle B$: $m\angle A + m\angle B = 90°$

$72.3° + m\angle B = 90°$

$m\angle B = 17.7°$

To find b: $\tan 72.3° = \dfrac{2.4}{b}$

$b\tan 72.3° = 2.4$

$b = \dfrac{2.4}{\tan 72.3°}$

$b \approx 0.765937...$

Then $b \approx 0.8$, correct to the nearest tenth.

To find c: $\sin 72.3° = \dfrac{2.4}{c}$

$c\sin 72.3° = 2.4$

$c = \dfrac{2.4}{\sin 72.3°}$

$c \approx 2.519257...$

Then $c \approx 2.5$, correct to the nearest tenth.

Thus, $m\angle B = 17.7°$, $b \approx 0.8$, $c \approx 2.5$.

17. Sketch a right triangle with $a = 7$ and $b = 11$.

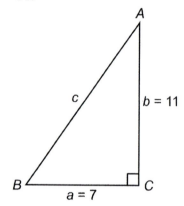

To find c: $c^2 = a^2 + b^2$

$c^2 = (7)^2 + (11)^2$

$c^2 = 49 + 121 = 170$

$c = \sqrt{170} \approx 13.038405...$

Then $c \approx 13$, correct to the nearest whole number.

To find $m\angle B$: $\tan B = \dfrac{11}{7}$

$\boxed{11}\ \boxed{÷}\ \boxed{7}\ \boxed{=}\ \boxed{\tan^{-1}} \rightarrow \boxed{57.528807...}$

Then $m\angle B \approx 58°$, correct to the nearest degree.

To find $m\angle A$: $\tan A = \dfrac{7}{11}$

$\boxed{7}\ \boxed{÷}\ \boxed{11}\ \boxed{=}\ \boxed{\tan^{-1}} \rightarrow \boxed{32.471192...}$

Then $m\angle A \approx 32°$, correct to the nearest degree.

Thus, $m\angle A \approx 32°$, $m\angle B \approx 58°$, $c \approx 13$.

18. Sketch a right triangle with $b = 12$ and $c = 20$.

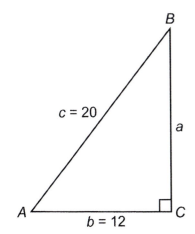

To find a: $c^2 = a^2 + b^2$

$$(20)^2 = a^2 + (12)^2$$

$$400 = a^2 + 144$$

$$256 = a^2$$

$$\sqrt{256} = a$$

$$16 = a$$

To find $m\angle A$: $\cos A = \dfrac{12}{20}$

$\boxed{12}\;\boxed{\div}\;\boxed{20}\;\boxed{=}\;\boxed{\cos^{-1}}\;\rightarrow\;\boxed{53.130102...}$

Then $m\angle A \approx 53°$, correct to the nearest degree.

To find $m\angle B$: $\sin B = \dfrac{12}{20}$

$\boxed{12}\;\boxed{\div}\;\boxed{20}\;\boxed{=}\;\boxed{\sin^{-1}}\;\rightarrow\;\boxed{36.869897...}$

Then $m\angle B \approx 37°$, correct to the nearest degree.

Thus, $m\angle A \approx 53°$, $m\angle B \approx 37°$, $a = 16$.

19. Sketch a right triangle with $a = 6.2$ and $c = 9.6$.

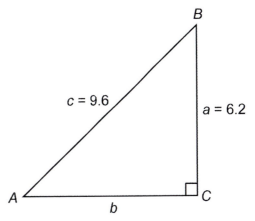

To find b: $c^2 = a^2 + b^2$

$$(9.6)^2 = (6.2)^2 + b^2$$

$$53.72 = b^2$$

$$\sqrt{53.72} = b$$

$$7.3293929 \approx b$$

Then $b \approx 7.3$, correct to the nearest tenth.

To find $m\angle A$: $\sin A = \dfrac{6.2}{9.6}$

$\boxed{6.2}\;\boxed{\div}\;\boxed{9.6}\;\boxed{=}\;\boxed{\sin^{-1}}\;\rightarrow\;\boxed{40.228184...}$

Then $m\angle A \approx 40.2°$, correct to the nearest tenth of a degree.

To find $m\angle B$: $\cos B = \dfrac{6.2}{9.6}$

$\boxed{6.2}\;\boxed{\div}\;\boxed{9.6}\;\boxed{=}\;\boxed{\cos^{-1}}\;\rightarrow\;\boxed{49.771815...}$

Then $m\angle B \approx 49.8°$, correct to the nearest tenth of a degree.

Thus, $m\angle A \approx 40.2°$, $m\angle B \approx 49.8°$, $b \approx 7.3$.

20. Sketch a rectangle with the given measurements.

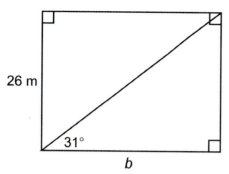

Let b = base of the rectangle.

To find the base b, use the tangent ratio.

$$\tan 31° = \frac{26}{b}$$
$$b \tan 31° = 26$$
$$b = \frac{26}{\tan 31°}$$

The formula for the area of a rectangle is $A = bh$.

$$A = \frac{26}{\tan 31°}(26)$$

$A \approx 1125.1$ m^2 (rounded to the nearest tenth of a square meter)

21. Sketch the information given.

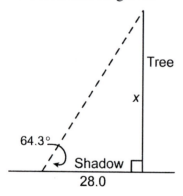

Let x be the height of the tree. Then

$$\tan 64.3° = \frac{x}{28.0}$$
$$28.0 \tan 64.3° = x$$
$$58.179702... \approx x$$

Then the tree is approximately 58.2 m tall, correct to the nearest tenth of a meter.

22. Sketch the information given.

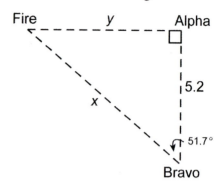

We must find x and y.

$$\cos 51.7° = \frac{5.2}{x}$$
$$x \cos 51.7° = 5.2$$
$$x = \frac{5.2}{\cos 51.7°}$$
$$x \approx 8.390087...$$

$$\tan 51.7° = \frac{y}{5.2}$$
$$5.2 \tan 51.7° = y$$
$$6.584342... \approx y$$

Thus, the fire is approximately 8.4 miles from Outpost Bravo and approximately 6.6 miles from Outpost Alpha.

23. Make a sketch using the information given.

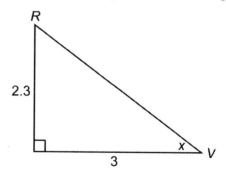

Let x = the angle of elevation.

$$\tan x = \frac{2.3}{3}$$

$$\boxed{\dfrac{2.3}{3}} \boxed{=} \boxed{\tan^{-1}} \rightarrow \boxed{37.476179...}$$

Thus the angle of elevation, to the nearest tenth of a degree is approximately 37.5°.

24. Make a sketch using the information given.

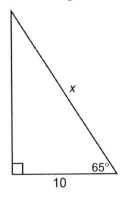

Let x = the length of the guy wire.

$$\cos 65° = \frac{10}{x}$$

$$x\cos 65° = 10$$

$$x = \frac{10}{\cos 65°} \approx 23.662015...$$

Thus the length of the guy wire, to the nearest tenth of a yard is approximately 23.7 yards.

25. Use the given sketch of the problem and let x = the distance traveled on the escalator.

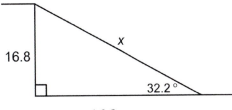

$$\sin 32.2° = \frac{16.8}{x}$$

$$x\sin 32.2° = 16.8$$

$$x = \frac{16.8}{\sin 32.2°} \approx 31.527018...$$

Thus the distance traveled, rounded to the nearest tenth of a foot is approximately 31.5 ft.

Chapter 10 Practice Test

1. Sketch a right triangle with $a = 5$ and $c = 9$.

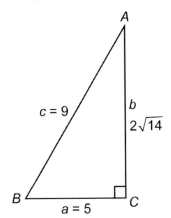

First use the Pythagorean Theorem to find b.

$$c^2 = a^2 + b^2$$

$$(9)^2 = (5)^2 + b^2$$

$$56 = b^2$$

Then $b = \sqrt{56} = \sqrt{4 \cdot 14} = 2\sqrt{14}$.

$$\sin A = \frac{a}{c} = \frac{5}{9}$$

$$\cos A = \frac{b}{c} = \frac{2\sqrt{14}}{9}$$

$$\tan A = \frac{a}{b} = \frac{5}{2\sqrt{14}} = \frac{5\sqrt{14}}{2\sqrt{14}\sqrt{14}}$$

$$= \frac{5\sqrt{14}}{2(14)} = \frac{5\sqrt{14}}{28}$$

2.

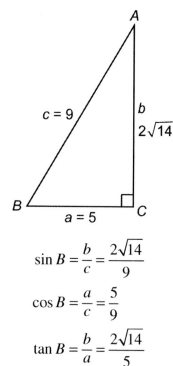

$$\sin B = \frac{b}{c} = \frac{2\sqrt{14}}{9}$$

$$\cos B = \frac{a}{c} = \frac{5}{9}$$

$$\tan B = \frac{b}{a} = \frac{2\sqrt{14}}{5}$$

3. Sketch a right triangle in which $\sin A = \frac{2}{5}$, that is, the side opposite $\angle A$ is 2 and the hypotenuse is 5.

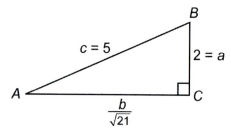

Use the Pythagorean Theorem to find b.

$$c^2 = a^2 + b^2$$

$$(5)^2 = (2)^2 + b^2$$

$$21 = b^2$$

$$\sqrt{21} = b$$

$$\cos A = \frac{b}{c} = \frac{\sqrt{21}}{5}$$

$$\tan A = \frac{a}{b} = \frac{2}{\sqrt{21}} = \frac{2}{\sqrt{21}} \cdot \frac{\sqrt{21}}{\sqrt{21}}$$

$$= \frac{2\sqrt{21}}{21}$$

4. $\boxed{68.9}\ \boxed{\sin} \rightarrow \boxed{0.932953...}$

Then $\sin 68.9° \approx 0.9330$, correct to four decimal places.

$\boxed{68.9}\ \boxed{\cos} \rightarrow \boxed{0.359996...}$

Then $\cos 68.9° \approx 0.3600$, correct to four decimal places.

$\boxed{68.9}\ \boxed{\tan} \rightarrow \boxed{2.591560...}$

Then $\tan 68.9° \approx 2.5916$, correct to four decimal places.

5. To find x, use the sine ratio.

$$\sin 47° = \frac{x}{21}$$

$$21 \sin 47° = x$$

Thus, x is approximately, rounded to the nearest tenth, 15.4 inches.

To find y, use the cosine ratio.

$$\cos 47° = \frac{y}{21}$$

$$21 \cos 47° = y$$

Thus, y is approximately, rounded to the nearest tenth, 14.3 inches.

6. To find x, use the cosine ratio.

$$\cos 39° = \frac{10}{x}$$

$$x \cos 39° = 10$$

$$x = \frac{10}{\cos 39°}$$

Thus, x is approximately, rounded to the nearest tenth, 12.9 mm.

To find y, use the tangent ratio.

$$\tan 39° = \frac{y}{10}$$

$$10 \tan 39° = y$$

Thus, y is approximately, rounded to the nearest tenth, 8.1 mm.

7. $\boxed{0.1096}\ \boxed{\cos^{-1}} \rightarrow \boxed{83.707742...}$

Thus, $m\angle A \approx 83.7°$, correct to the nearest tenth of a degree.

8. Sketch the right triangle with $a = 1.1$ and $c = 4.5$.

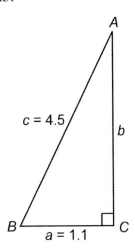

To find b: $\quad c^2 = a^2 + b^2$

$$(4.5)^2 = (1.1)^2 + b^2$$

$$19.04 = b^2$$

$$\sqrt{19.04} = b$$

Then $b \approx 4.363484...$ so $b \approx 4.4$, correct to the nearest tenth.

To find $m\angle A$: $\ \sin A = \dfrac{1.1}{4.5}$

$$m\angle A \approx 14.149004...$$

$\boxed{1.1}\ \boxed{\div}\ \boxed{4.5}\ \boxed{=}\ \boxed{\sin^{-1}} \rightarrow \boxed{14.149004...}$

Then $m\angle A \approx 14.1°$, to the nearest tenth.

To find $m\angle B$: $\cos B = \dfrac{1.1}{4.5}$

$\boxed{1.1}\ \boxed{\div}\ \boxed{4.5}\ \boxed{=}\ \boxed{\cos^{-1}} \rightarrow \boxed{75.850995...}$

Then $m\angle B \approx 75.9°$, to the nearest tenth.

Thus, $m\angle A \approx 14.1°$, $m\angle B \approx 75.9°$, and $b \approx 4.4$.

9. Sketch the right triangle with $b = 15$ and $m\angle A = 38°$.

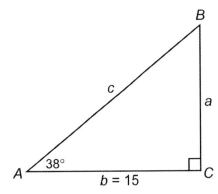

To find $m\angle B$: $\quad m\angle A + m\angle B = 90°$

$$38° + m\angle B = 90°$$

$$m\angle B = 52°$$

To find a: $\qquad \tan 38° = \dfrac{a}{15}$

$$15 \tan 38° = a$$

$$11.719284... \approx a$$

Then $a \approx 12$, correct to the nearest whole number.

To find c: $\quad \cos 38° = \dfrac{15}{c}$

$$c \cos 38° = 15$$

$$c = \dfrac{15}{\cos 38°}$$

$$c \approx 19.035273...$$

Then $c \approx 19$, correct to the nearest whole number.

Thus $m\angle B = 52°$, $a \approx 12$, $c \approx 19$.

10. The formula for the area of a rectangle is $A = bh$.

Let b = the base of the rectangle and h = the height of the rectangle. (See the figure.)

Find b by using the cosine ratio.

$$\cos 30° = \frac{b}{6.2}$$

$$6.2 \cos 30° = b$$

Find h by using the sine ratio.

$$\sin 30° = \frac{h}{6.2}$$

$$6.2 \sin 30° = h$$

The area of the rectangle is

$$A = (6.2 \cos 30°)(6.2 \sin 30°)$$

Thus, the area is approximately, rounded to the nearest tenth, 16.6 cm^2.

11. Use the given sketch of the problem.

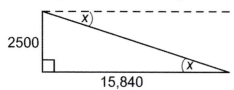

Let x = the needed angle of glide. (In the figure, alternate interior angles are marked congruent.)

Remember 5280 feet = 1 mile.

The distance to the airport is 3(5280) = 15,840 ft.

$$\tan x° = \frac{2500}{15,840}$$

$$\boxed{\frac{2500}{15,840}} \boxed{=} \boxed{\tan^{-1}} \rightarrow \boxed{8.968911...}$$

Thus the angle of the glide, rounded to the nearest degree is approximately 9°.

12. Make a sketch using the given information.

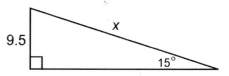

Let x = the length of the ramp.

$$\sin 15° = \frac{9.5}{x}$$

$$x \sin 15° = 9.5$$

$$x = \frac{9.5}{\sin 15°} = 36.705181$$

Thus rounded to the nearest tenth of an inch, the ramp should approximately 36.7 inches long.

CHAPTERS 8-10 CUMULATIVE REVIEW

1. The formula for the total surface area of a right prism is $SA = LA + 2B$ where $LA = ph$.

$$p = 2(5.4) + 2(2.3) = 15.4 \text{ cm}$$

$$LA = (15.4)(3.4) = 52.36 \text{ cm}^2$$

$$B = (5.4)(2.3) = 12.42 \text{ cm}^2$$

$$SA = 52.36 + 2(12.42) = 77.2 \text{ cm}^2$$

$$V = Bh$$

$$V = (12.42)(3.4) = 42.228 \text{ cm}^3$$

2. The formula for the total surface area of a regular square pyramid is $SA = LA + B$ where $LA = \frac{1}{2}p\ell$.

$$p = 4(6) = 24 \text{ ft}$$

To find the slant height, use the Pythagorean Theorem.

$$\ell^2 = 3^2 + 4^2$$

$$\ell^2 = 9 + 16$$

$$\ell^2 = 25$$

$$\ell = 5$$

$$LA = \frac{1}{2}(24)(5) = 60 \text{ ft}^2$$

$$B = (6)^2 = 36 \text{ ft}^2$$

$$SA = 60 + 36 = 96 \text{ ft}^2$$

$$V = \frac{1}{3}Bh$$

$$V = \frac{1}{3}(36)(4) = 48 \text{ ft}^3$$

3. Use the volume formula for a right cylinder, $V = Bh$.

$d = 2r = 2.5$. Thus $r = 1.25$ in.

$$B = \pi r^2 = \pi(1.25)^2$$

$$V = \pi(1.25)^2(5) \approx 24.543692...$$

Thus the volume of the can, rounded to the nearest tenth is approximately 24.5 in^3.

4. The formula for the surface area of a square pyramid is $SA = LA + B$ where $LA = \frac{1}{2}p\ell$.

Since the edge of the square is 12 in, the perimeter is 48 in.

Use the Pythagorean Theorem to find the slant height.

$$\ell^2 = 6^2 + 8^2$$

$$\ell^2 = 100$$

$$\ell = 10$$

$$LA = \frac{1}{2}(48)(10) = 240 \text{ in}^2$$

$$B = (12)^2 = 144 \text{ in}^2$$

$$SA = 240 + 144 = 384$$

Thus the surface area of the display pyramid is 384 in^2.

5. Find the volume of the ice cream using the formula $V = \frac{4}{3}\pi r^3$.

$$V_{sphere} = \frac{4}{3}\pi(1)^3 \approx 4.188790...$$

Thus the volume of the sphere is about 4.1 in^3.

Find the volume of the cone using the formula, $V = \frac{1}{3}Bh$.

The radius of the cone is 1 inch.

$$B = \pi(1)^2$$

$$V_{cone} = \frac{1}{3}\pi(1)^2(4.5) \approx 4.712388...$$

Thus the volume of the cone is about 4.7 in^3.

Since the volume of the cone is larger than the volume of the ice cream, the cone will hold all the ice cream when it melts.

6.

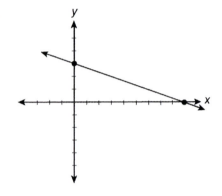

To find the x-intercept, substitute 0 for y and solve for x.

$$x + 3(0) = 9$$

$$x = 9$$

Thus, the x-intercept is $(9, 0)$.

To find the y-intercept, substitute 0 for x and solve for y.

$$0 + 3y = 9$$

$$y = 3$$

Thus, the y-intercept is $(0, 3)$.

To find the distance between the two intercepts, use the distance formula:

$$d = \sqrt{(x_2 - x_1)^2 + (y_2 - y_1)^2}$$

$$d = \sqrt{(9-0)^2 + (0-3)^2}$$

$$d = \sqrt{81 + 9}$$

$$d = \sqrt{90} \text{ or } 3\sqrt{10}$$

7. To find the slope of the line, use the slope formula.

$$m = \frac{y_2 - y_1}{x_2 - x_1} = \frac{5 - (-3)}{-2 - 0} = \frac{8}{-2} = -4$$

Since the point $(0, -3)$ is the y-intercept, put the equation into slope-intercept form: $y = mx + b$.

$$y = -4x - 3$$

8. Let $\overline{AB}$ be the given segment with endpoints $(-a, 0)$ and $(a, 0)$ then $(0, 0)$ is the midpoint of the segment. The perpendicular bisector is the y-axis. Choose any point C on the y-axis, call it $(0, b)$. Prove $AC = BC$. (See the figure.)

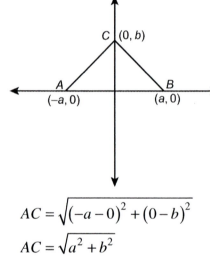

$$AC = \sqrt{(-a - 0)^2 + (0 - b)^2}$$

$$AC = \sqrt{a^2 + b^2}$$

$$BC = \sqrt{(a - 0)^2 + (0 - b)^2}$$

$$BC = \sqrt{a^2 + b^2}$$

Thus $AC = BC$ and any point on the perpendicular bisector of a segment is equidistant from the endpoints of the segment.

9. Use the information given on the figure.

To find a, use the cosine ratio.

$$\cos 41° = \frac{\text{adjacent leg}}{\text{hypotenuse}} = \frac{10.2}{a}$$

$$a \cos 41° = 10.2$$

$$a = \frac{10.2}{\cos 41°} \approx 13.515132...$$

To find b, use the tangent ratio.

$$\tan 41° = \frac{\text{opposite leg}}{\text{adjacent leg}} = \frac{b}{10.2}$$

$$10.2 \tan 41° = b$$

$$b \approx 8.866724...$$

Thus, to the nearest tenth, $a \approx 13.5$ and $b \approx 8.9$

10. Use the information given on the figure.

To find $m\angle B$, use the sine ratio.

$$\sin B = \frac{\text{opposite leg}}{\text{hypotenuse}} = \frac{12}{13}$$

Use the following keystrokes on a scientific calculator to find $m\angle ABC$:

$$\boxed{\frac{12}{13}} \; \boxed{=} \; \boxed{\sin^{-1}} \rightarrow \boxed{67.380135...}$$

Thus, the $m\angle ABC \approx 67°$ measured to the nearest degree.

11. Sketch a right triangle with $a = 2.7$, $b = 3.4$, and $m\angle C = 90°$.

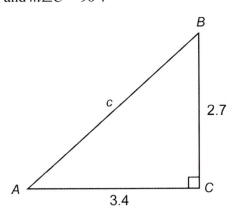

To find c, use the Pythagorean Theorem.

$$c^2 = a^2 + b^2$$

$$c^2 = (2.7)^2 + (3.4)^2$$

$$c^2 = 18.85$$

$$c = \sqrt{18.85} \approx 4.341658...$$

$$c \approx 4.3 \text{ (rounded to the nearest tenth)}$$

To find $m\angle A$, use the tangent ratio.

$$\tan A = \frac{\text{opposite leg}}{\text{adjacent leg}} = \frac{2.7}{3.4}$$

Use the following keystrokes on a scientific calculator to find $m\angle A$:

$$\boxed{\frac{2.7}{3.4}} \; \boxed{=} \; \boxed{\tan^{-1}} \rightarrow 38.453709...$$

Thus $m\angle A \approx 38.5°$ (rounded to the nearest tenth of a degree).

To find $m\angle B$, use the tangent ratio.

$$\tan B = \frac{\text{opposite leg}}{\text{hypotenuse}} = \frac{3.4}{2.7}$$

Use the following keystrokes on a scientific calculator to find $m\angle B$:

$$\boxed{\frac{3.4}{2.7}} \; \boxed{=} \; \boxed{\tan^{-1}} \rightarrow 51.546290...$$

Thus $m\angle B \approx 51.5°$ (rounded to the nearest tenth of a degree).

Thus $m\angle A \approx 38.5°$, $m\angle B \approx 51.5°$, $c \approx 4.3$.

12. Sketch a figure using the given information.

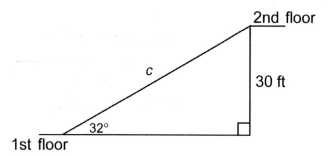

Let c = the length of the escalator.
To find c, use the sine ratio.

$$\sin 32° = \frac{\text{opposite leg}}{\text{hypotenuse}} = \frac{30}{c}$$

$$c \sin 32° = 30$$

$$c = \frac{30}{\sin 32°} \approx 56.612397...$$

Thus, the length of the escalator, rounded to the nearest foot, is approximately 56.6 ft.